Mothers
Who
Can't Love

Mothers Who Can't Love

A Healing Guide *for* Daughters

Susan Forward, PhD

WITH DONNA FRAZIER GLYNN

HARPER

www.harpercollins.com

HarperCollins books may be purchased for educational, business, or sales promotional use. For information, please e-mail the Special Markets Department at SPsales@harpercollins.com.

FIRST EDITION

Designed by Fritz Metsch

Library of Congress Cataloging-in-Publication Data has been applied for.

ISBN: 978-0-06-220434-9

13 14 15 16 17   OV/RRD   10 9 8 7 6 5 4 3 2 1

To my treasured daughter, Wendy

Contents

❦

Mothers
Who
Can't Love

Introduction

"I was on a business trip to Wisconsin. I had been cooped up inside all day and I wanted to get some air. The sun was out, so at the lunch break, even though it was pretty cold outside, I decided to take a short walk. I looked for the sunniest spot I could find, but you know it was the damnedest thing—it sure looked like the sun and it was bright like the sun, but there was absolutely no warmth coming from it. And this wave of sadness came over me—the sun was just like my mother."

Heather, a petite, thirty-four-year-old sales rep for a large pharmaceutical company, became teary as she spoke. She was expecting her first child and was frightened that she might be at risk of becoming the same kind of mother she'd had.

HEATHER: "You know, for the longest time, I couldn't even think about being a mom myself. I felt so lucky when I met Jim after a series of bad relationships and I realized that someone could really love me. We've wanted to have a baby for a long time, but I was so afraid there was something wrong with me. Like maybe all the coldness from my mom would start coming out of me once I was pregnant. I couldn't bear the thought that I might ever be that way with my own child."

It's the kind of upsetting story I've heard again and again from women who carry with them a legacy of pain, fear, and turmoil because of the profound emotional wounds inflicted by their mothers.

In more than thirty-five years as a therapist working in a variety of clinical settings, I've seen large numbers of women like Heather, who, knowingly or unknowingly, are caught in the damaging emotional orbit of the women who brought them up and are struggling to escape. They come to therapy sessions with anxiety and depression, relationship problems, lack of confidence, concerns about their ability to stand up for themselves, or even to love. Some are able to make a connection between their relationship with their mothers and the difficulties in their lives. Others mention, "My mother is driving me crazy," but consider that to be secondary to the issues that bring them to me.

Often they're sorting through confusing mixed messages, hoping to prove themselves wrong about the pain they carry from the past.

I needed to hear more about the fears Heather was carrying into motherhood, so I asked her to tell me what she meant specifically by "the coldness from my mom" that she was so afraid of replicating with her own child. She began hesitantly:

HEATHER: "It was like my mom had two sides—she gave me birthday parties, sometimes she came to events at school—she could even be nice to my friends. But then she had this other side. . . ."

"And what was that like?" I asked.

HEATHER: "Well, she criticized me an awful lot—but to tell you the truth, most of the time she ignored me, like I wasn't even worth her time. I don't know—maybe the nice things she did were all for show. But I'll tell you, I sure didn't get to feel safe around her—there was no real bond or kindness. . . . I never felt important to her. I was just something she had to deal with when it suited her. But she was busy. You can't blame a single mom for being distracted."

Like so many women, Heather could speak candidly about how she'd been treated. Yet she grasped for ways to minimize the hurt and struggled to see her mother as something she'd rarely been: loving.

What Makes a Good Mother?

A good mother is not expected to be perfect and self-sacrificing to the point of martyrdom. She has her own emotional baggage, her own scars, her own needs. She may have work that she doesn't want to compromise, and there may be times when she's not available to her daughter. She may lose her temper, and say or do things to her daughter that she regrets. But if her *dominant* behavior engenders in her daughter a belief in her own value and nourishes her self-respect, confidence, and safety, that mother is doing a good job, whether she's a wonderful mom or just good enough. She's demonstrating real love, in a tangible, reliable way, to her child.

That's not the kind of mothering Heather, and so many other women, experienced. For them, nourishing love and attention always came drop by drop. Behind closed doors, those intermittent splashes of warmth inevitably gave way to a reality that outsiders rarely saw: Their mothers tore them down, competed with them, icily ignored them, took credit for their achievements, failed to protect them, or even abused them. But love them? No. Loving is consistent overall behavior, and daughters like Heather were starved for its nurturing warmth.

The High Cost of Missing
a Mother's Love

The effects of growing up this way are painful and wounding. Girls define their emerging womanhood by identifying and bonding with their moms. But when that vital process is distorted—because their mothers are abusive, critical, smothering, depressed, neglectful, or distant—they're left to struggle alone to try to find a solid sense of themselves and their place in the world.

It rarely occurs to them that their mothers were not loving, or even, in extreme cases, that they were malevolent. That's too hard to admit, and allowing in that possibility produces acute anxiety in children, whose survival is so closely tied to their vital caretaker. It's far safer for a child to believe that "if there's something wrong between us, it's because there's something wrong with *me*." She makes sense of her mother's hurtful behavior by turning it into self-blame and feelings of inadequacy and badness, feelings that persist into adulthood no matter how accomplished she is or how much she is loved by others, including her own children.

A little girl who was criticized or ignored or abused or stifled by an unloving mother becomes an adult who tells herself she'll never be good enough or lovable enough, never smart or pretty or acceptable enough to deserve success and happiness. *Because if you really were worthy of respect and affection,* a voice inside whispers, *your mother would've given them to you.*

If you were that little girl, the daughter of a mother who couldn't give you the love you needed so much, it's likely that much like Heather, you now go through your days with a cavernous gap in your confidence, a sense of emptiness and sadness. You're never truly comfortable in your own skin. You may not trust your ability to love. And you can't step fully into your life until you heal that gaping mother wound.

Why I'm Writing This Book Now

My session with Heather brought that painful reality home to me once again, and I thought about her long after our session was over. She was intelligent, attractive, and accomplished, yet those qualities seemed largely invisible to her. She doubted her ability to love and be loved, and, I'd discover, she felt like a fraud, ever fearful that there was something wrong with her, despite all evidence to the contrary. Self-aware as she was, at age thirty-four she was still waiting for a mother's acknowledgment and blessing to give her confidence, faith in herself as a woman, a partner, a mother. And it would probably never come. When a strong mother bond is missing, women often struggle for a lifetime with a bewildering sense of loss and a sense of deprivation.

I've always specialized in shining a light on the difficult truths of how we really treat one another, beneath the facades of "the perfect couple" or "the happy family," and after writing *Toxic Parents*, I thought I'd said all I had to say about the people who raise us. But as more and more daughters kept bringing their mother wounds to me, I knew I needed to speak woman-to-woman to the millions who are grappling with an unloving mother and the soul-deep consequences of having her in their lives.

There was another—quite telling—factor in my decision to write this book. Though I had long resolved the turmoil caused by my own mother, until she died I couldn't give myself permission to write a book focusing on mothers who can't love. My women clients often have had painful relationships with their fathers as well— fathers who had massive problems of their own and were rarely available for their daughters, since a healthy man usually doesn't marry or stay with an unstable woman. But the jagged relationship with their mothers always seems to be at the emotional core as daughters try to navigate marriages, careers, and motherhood.

If you've lived with an unloving mother, her legacy is there every day in the difficulties you face, and keep facing, in your emotional relationships and as you try to develop confidence and self-respect. I know you may be frustrated, discouraged, and confused. But I want you to know that we'll work together in this book to find the clarity and relief that have eluded you. I'll guide you step by step in reshaping your relationship with your mother and yourself to heal the wounds that have brought so much pain for so long.

Starting now, we'll take a close, honest look at your mother's behavior and the impact it's had on you. As we move through this book, you'll see her patterns—and yours—in detail. I'll give you effective new strategies for changing the beliefs and behaviors that have held you back. And I'll help you gain an understanding, perhaps for the first time, of what real love from a parent, or anyone, looks and feels like. That's the powerful, reliable touchstone that will guide you in rebuilding your life.

You Can't Call It Love

To help you look objectively at the mothering you've experienced, I've designed the following checklists. First, let's deal with what's going on right now.

Does your mother regularly:

- Demean or criticize you?
- Make you a scapegoat?
- Take credit when things go well, and blame you when they go wrong?
- Treat you as if you're incapable of making your own decisions?
- Turn on the charm for other people, but turn cold when she's alone with you?
- Try to upstage you?

- Flirt with your significant other?
- Try to live out her life through you?
- Call, e-mail, text, and schedule herself into your life so much that you feel smothered?
- Tell you or imply that you are the reason for her depression, lack of success, or unfulfilled life?
- Tell you or imply that she can't cope without you (and only *your* help will do)?
- Use money or the promise of money to manipulate you?
- Threaten to make your life difficult if you don't do what she wants?
- Ignore or discount your feelings and wants?

"Yes" answers are clear indications that your mother is crossing, or has crossed, the boundary that separates loving mothers from unloving ones. These behaviors are probably not new, and chances are they have been going on for most of your life. You'll see that clearly if you put a simple "Did she" in front of each of the questions above, and think about what was going on when you were little.

The next list will give you a sense of how your relationship with your mother has affected you.

Do you:

- Wonder if your mother loves you—and feel ashamed that she may not?
- Feel responsible for the happiness of everyone but yourself?
- Believe that your mother's needs, wants, and expectations of you are more important than your own?
- Believe that love is something you have to earn?
- Believe that no matter what you do for your mother, it won't be enough?
- Believe that you must protect her, even from the knowledge that she's hurting you?

- Feel guilty and believe you're a bad person if you don't comply with the wishes of other people, especially your mother?
- Hide the details of your life and feelings from your mother, because you know she'll find a way to use your truths against you?
- Find yourself constantly chasing approval?
- Feel scared, guilty, and small, no matter how much you accomplish?
- Wonder if there's something wrong with you that will keep you from being able to find a partner who loves you?
- Feel afraid to have children (if you want them) because they'll turn out "messed up, like me"?

All these feelings and beliefs are legacies of the mother wound, and they, too, have their roots in childhood. But even if you answered "yes" to every one of these questions, please be assured that you're not doomed or irreparably damaged. There are many changes you can put into practice immediately to improve your life, your self-image, and your relationships.

The women you'll meet in this book are very much like you. Throughout the pages that follow, you'll see how they courageously examine their pasts and use their new understanding of their mothers, and themselves, to make enormous positive changes in their lives. I'll guide you through the same healing journey that unfolds in our counseling sessions, giving you the tools to finally free yourself from the painful legacy of growing up with an unloving mother.

How This Book Is Organized

The chapters in the first half of the book will introduce you to the five common varieties of mothers who can't love. I'll take you into therapy sessions with daughters like Heather, and you'll see each type of mother's behavior through her daughter's eyes. You'll have a chance to listen as daughters describe the difficulties they face now with their mothers and how that affects the rest of their lives. You'll also see the coping mechanisms they used as girls to protect themselves from their mother's unloving behavior, and you'll see how those behaviors and beliefs have crystallized into some very painful, self-defeating patterns.

You'll be able to look at behavior like your mother's in a context stripped of the explanations and rationalizations you've probably heard all your life, and you'll come to understand your mother much more clearly. I'm sure you'll see yourself more clearly as well. It's quite likely that you'll see elements of your own experiences reflected in more than one chapter—many unloving mothers fall into more than one category, and daughters who have been deprived of healthy mothering, no matter what type, carry many of the same scars.

In the second half of the book, we move from discovery to recovery. I'll guide you through the steps and strategies that will allow you to change your relationship with your mother and improve your life. We'll work together to move the understandings in your head into the emotional realm, where you'll be able to make deep shifts in the way you see both your mother and yourself. Then I will introduce you to the tools you can use to reclaim your confidence, your self-esteem, and your fragile sense of lovability.

What I can tell you, from my own experience and that of thousands of other daughters, is that it's possible to change the belief you've carried so long that something in you shattered when you were young and can't be mended. I promise you that as we work together in this book you will gain a wonderful sense of wholeness. In yourself and the world around you, you will find paths to the self-respect, wisdom, and caring that you've craved for so long.

Part One

⚘

Identifying
the
Mother Wound

Chapter 1

The Taboo of Questioning Your Mother's Love

᭢

"Don't you dare say anything bad about your mother."

W e may think we live in very psychologically aware times, but we haven't yet managed to shake off our mythical version of motherhood—the myth that says a mother by definition is capable of love, protection, and kindness. The mother myth gives great cover to unloving mothers, who far too often operate undisturbed while their husbands, other family members, and society deflect any criticism or scrutiny aimed at them.

Most societies glorify mothers, as if the mere act of giving birth makes them inherently capable of nurturing. That's simply not true. There's no magic switch that turns on "maternal instinct" and ensures that a woman, especially a troubled one, will suddenly bond with her baby, know and respond to what that child needs, and give her the nurturing she craves. Of course the Freudian tradition of mother-bashing—blaming mothers for everything that goes wrong—is erroneous, but it is also a fantasy to believe that the role of mother is automatically synonymous with healthy love.

The widespread belief in this fantasy is so strong that if

your mother was unloving and you try to tell the truth of what happened—how your mother *really* behaved toward you—you'll inevitably run into a wall of resistance from powerful external forces (including your mother herself) that pull together to defend her.

In fact, dealings with unloving mothers are constricted by so many taboos, and attitudes about the mother wound are so charged, that it's common to encounter skepticism, sharp criticism, and counterproductive advice. If you've attempted to chart a new course with your mother, you've already seen what can happen:

- You try to make peace with her and find yourself pulled back into a web of criticism and manipulation. Once more you're the ungrateful one. The selfish one. The unforgiving one. The one who will always owe her, no matter what she does.
- You seek advice from relatives and sometimes from friends, who respond, "How can you talk that way about your mother? She gave you life. What's *wrong* with you?"
- You have the misfortune of consulting misguided therapists who urge you to "forgive and forget" and make peace with your mother, no matter how high the emotional cost to you.
- You try to get support from a priest, minister, or spiritual counselor and you're met with responses like: "Honor thy mother." "You won't heal until you forgive." "Family is everything."
- You even try talking to your partner, who counsels: "Don't let her get to you. That's just the way she is."

And after all that, you're back where you started—bewildered, alone, and even shamed by the attempt to face and overcome your history. You may start to wonder if you have the right to feel the way you do.

Other People Don't See What You See

Struggling with the pain and repercussions of having an unloving mother can be intensely lonely and isolating. People with reasonably healthy mothers have a tough time understanding that all mothers are not like theirs, and it's quite common for even a well-meaning friend or relative to discount an unloved daughter's pain or blame her when she looks for sympathy. My client Valerie, a thirty-two-year-old computer programmer, came to me hoping to overcome the shyness and anxiety that had made her feel stuck and discouraged about her work and social life. It was hard to break out of her shell, she said, especially because "people just don't understand me."

When I asked for an example, she described a recent incident:

VALERIE: "A month ago, I enrolled in an adult art class—
    something I've always wanted to do. The instructor, Terry,
    was very encouraging about my watercolors, and despite a
    twenty-five-year difference in our ages, we became quite
    close. Terry told me that she was going to put on an exhibit
    of student work, and I was thrilled when she told me she had
    selected two of my pieces. Suddenly I burst into tears. When
    she asked me what was going on, I told her that I'd had a
    huge fight with my mother on the phone that morning and I
    didn't want to invite her.

    "She said she was eager to meet my mother, who's an
    interior designer—and also a frustrated artist, though I didn't
    mention that. Terry brought up my mom a couple of times,
    so against my better judgment I sent her an e-mail. So Mom
    comes to the exhibit and praises everyone else's work, and she's
    totally lukewarm about mine. But, of course, she was her most
    charming and effusive to Terry. After she left, Terry said to me,

'I would love to have a lovely mother like that—I would give anything if my mother were still alive. I hope you know how lucky you are.'

"I said, 'Well—what you see isn't always what you get. My mother can be very self-centered, critical, and competitive.' It was as if Terry didn't hear a word I said. She just repeated: 'You should thank your lucky stars you have a mother who cares enough about you to come to see your work.'"

Valerie felt frustrated and unheard by a woman she thought was her friend. And if the pain is deep for an adult like Valerie, imagine how much worse it was for a young, dependent girl—or your own younger self—as she tried to feel heard and understood.

Colleen, a single, twenty-eight-year-old manager for a supermarket chain, told me that for as long as she could remember, she had suffered from a chronic, low-grade depression. She was on medication that had helped her considerably, but she had enough insight to see that there was a lot of unfinished business in her past that was contributing to her depressed feelings, and she wanted my help to sort through it. After I found out a little more about her, I asked her to tell me something about her childhood:

COLLEEN: "I had no one to talk to. . . . It was no bed of roses. Nobody listened to me, and I just stuffed my sadness. If I tried to talk to my father about my mother, he would say, 'Just be nice to her.' Once I got to stay overnight with my aunt Gina. She asked me how things were at home. I felt pretty safe with her—she'd always been okay with me—so I told her, 'I think something's wrong with Mom. She's always screaming at me and telling me I'm not worth anything.' Gina listened quietly and seemed to understand, but then she said, 'You have to try to keep her happy—she doesn't mean those things. She's very unhappy with your father, and if it wasn't for you, she would have left him a long time ago. You owe her. Don't be so sensitive.' She sounded really upset with me. I felt even worse after I told her—and I thought, 'Great! Now she's mad at me too.' I wished I'd never said anything."

The great common denominator among women with unloving mothers is the longing for validation—to find someone who will say, "Yes, what you experienced really happened. Yes, your feelings are justified. I understand."

Great pressure is brought to bear on daughters to not tell about the verbal, emotional, and even physical cruelties of their past and present. As you can see, for children the rules become clear early: *Don't tell anyone. Don't even tell yourself.*

That's how you learn to bury, minimize, and mistrust your own truth.

How Maternal Rejection Is Internalized

That impulse to make the best of what you have where your mother is concerned may seem positive, but it masks a whole system of fault lines that run below the surface of your life, a sort of emotional earthquake zone. Maintaining the "grin and bear it" status quo keeps the peace, but it's less a choice than a form of paralysis brought about by shame and fear. What keeps you in the dark is a neat bit of emotional alchemy and amnesia.

All the external messages you've gotten whenever you've tried to tell the truth about your mother echo back at you from the inside as powerful emotions:

• You feel tremendous disloyalty for "criticizing her" by telling the truth. "After all, she gave you life."

- You feel ashamed. "All mothers love their children—and they must have a good reason if they don't."
- You doubt your perceptions and wonder if you're "too sensitive" or just feeling sorry for yourself.

These thoughts and feelings are intense, and for many of us they're so frightening, and touch such a deep well of pain and insecurity, that they spark what I can only describe as terror. It is terror of the consequences of allowing ourselves to experience what we'd feel if we admitted to ourselves that our mothers were unloving and of trying to change the relationship.

These are the words I so often hear daughters use to explain why they couldn't possibly label their mothers, even the most abusive of them, as unloving:

- I couldn't stand the guilt.
- I couldn't bear the sadness.
- I couldn't stand the loss.

The frightened little child inside the adult woman says, "If you tell the truth, it could mean you don't have a mother anymore." And when they hear the whisper of that inner child, even the most accomplished and sophisticated women forget that they're adults who no longer need close ties to their mothers to survive.

Once you've persuaded yourself you couldn't stand the feelings that come with acting on your truth, there's just one recourse: rationalizations that distort both your own self-image and your view of your mother.

"Listen, she really had it tough," you tell yourself. "I've got to cut her some slack."

Colleen struggled mightily to hold on to any shred of evidence that her mother "wasn't that bad," and her rationalizations were all too familiar.

COLLEEN: "I don't want you to think I demonize my mom—I don't. I mean, she made sure we had food and a roof over our heads—I never went hungry. I had books for school, nice clothes. And to be totally honest, I was kind of a troublemaker when I was little. It's no wonder she got upset with me."

Colleen was still hoping to somehow salvage something positive from a relationship with a mother who may have given her enough to eat, but who starved her emotionally. Doing that, though, required her to pull out the self-blame that's so familiar, and so oddly comforting, to daughters of unloving mothers.

Do you see the cycle at work here? The pain in your relationship with your mother keeps turning to fear, which becomes rationalization and self-blame. It's a closed loop that keeps you locked in and unable to change. Our intellects may know what's going on, but our emotions tell us a different story—and most of the time it's our emotions that we listen to.

Daughters of unloving mothers are often able to say: "My mother is depressed." "My mother is incredibly self-absorbed." "My mother is driving me crazy." And even "My mother is an alcoholic," "My mother was verbally abusive, and still is," or "My mother is such a bitch." The words sound tough and knowing, but it's a kind of knowing that often stops far short of giving you any relief. Because until you disconnect fully and completely from the mother myth, you can't halt the emotional cycle that is programmed to explain everything in just one way: "Whatever my mother did, it was all my fault."

All your life, you've probably been trapped in the belief that you, not your mother, are flawed. This damaged self-image shaped your developing sense of yourself as a woman, which you carried into adulthood like a steamer trunk. And from that early collection of fears and misunderstandings about yourself, you continue to orchestrate many of the self-defeating behaviors of your life.

Confronting the Taboos

This book will give you a detailed portrait of unloving mother-hood that will help you bury the mother myth once and for all. In the chapters that follow, you'll see a wide variety of mothers who, because of serious psychological or physiological impairment, are neither willing nor able to provide the kind of consistent love that plays such a large role in guiding a child toward emotional well-being. They cannot truly love.

I want to emphasize that none of these mothers wakes up in the morning thinking, "How can I hurt my daughter today?" Much of their behavior is driven by forces outside their conscious aware-ness, or emotions they are afraid of confronting: a crippling sense of insecurity, an unshakable feeling of deprivation, deep disappoint-ment in their own lives. As they look for relief from their own fears and sadness, they use their daughters to shore up their feelings of power or agency or control. The hallmark of all these mothers is a lack of empathy, and their intense self-centeredness blinds them to the suffering they create. They rarely step out of themselves to see things from your point of view. All they know is that they want what they want, and need what they need, and they find it difficult, if not impossible, to make the connection between their demons and the hurtful actions that come to define their relationship with you.

Please don't look away when the mothers' behavior you see in these pages seems painfully close to what you experienced. It's im-portant to recognize it as the opposite of love, and to let that recog-nition in deeply and fully, even if you have to do it a little at a time. I know it may be difficult, but we can't repair the damage wrought by the mother myth without taking a clear-eyed look at what your mother did, and the marks it left on your life.

The mothers you'll see in this half of the book fall into five recognizable types. There are no firm boundaries among these categories, and an unloving mother may fall into several. You will come to know the workings of:

- THE SEVERELY NARCISSISTIC MOTHER. Powerfully insecure and self-absorbed, she has an insatiable need for admiration and a grandiose sense of her own importance. She must be the center of attention and lunges for the spotlight anytime she feels it moving from herself to you. She may treat her daughter as a rival, undercutting her sense of confidence, attractiveness, and power as a woman. Criticism and competition flare anytime this mother feels threatened—particularly when her adult daughter begins to thrive.

- THE OVERLY ENMESHED MOTHER smothers her daughter with demands for time and attention, erasing the boundaries between them and insisting on being the most important person in her daughter's life—no matter what the cost. Because she relies on her role as a mother to fill all her emotional needs, she can't foster her daughter's healthy independence. She commonly describes her daughter as her "best friend," though she rarely empathizes when her daughter's needs and preferences don't line up with her own.

- THE CONTROL FREAK feels powerless in many parts of her life and uses her daughter to fill that void, seeing her as a person whose role in life is to make her mother happy and do her bidding. These mothers make their needs, wants, and demands clearly known, and threaten severe consequences anytime their daughters try to honor a different agenda. They justify their actions by insisting that only they know the best course of action their daughters can take, and their constant criticism makes daughters believe it.

- MOTHERS WHO NEED MOTHERING are overwhelmed. Often caught in the undertow of depression or addiction, they leave their daughters in the position of having to care for them, and often the rest of the family as well. Classic patterns of role reversal take hold as the daughter is thrust out of her own childhood to parent her childlike mother, all the while starved for the guidance and protection her mother is unable to give her.

- MOTHERS WHO NEGLECT, BETRAY, AND BATTER. These mothers occupy the darkest end of the spectrum, icily unable to summon any warmth at all, leaving their daughters unprotected from abuse at the hands of other family members—or even physically abusing their daughters themselves. The damage they inflict is poisonous, and the scars their daughters bear are deep.

We will see how all of these mothers chip away at the foundation of their daughters' lives, and through these examples, you will begin to understand how living with your own unloving mother taught you ways of being in the world that have impaired your ability to love, trust, and thrive.

Repeating Patterns

It's an old cliché that women tend to marry their fathers, but the more eye-opening truth is that we often marry our mothers. That is, in choosing partners and situations in adult life, we are frequently propelled by a strong unconscious need to repeat the familiar dramas that produced the mother wound.

As you saw earlier in the checklists, problematic patterns of caretaking, people-pleasing, and insecurity often have their roots in this relationship. From an unloving mother, a girl develops high tolerance for mistreatment, and at the darkest end, a battered child may become a battered adult or an abusive mother herself. But whatever your mother's legacy, the links you make between past and present will give you the desire, and the power, to make lasting changes.

Chapter 2

The Severely Narcissistic Mother

ᴠ

"But what about *me?*"

According to ancient Greek legend, there once lived a handsome young man named Narcissus who was so beautiful that both men and women fell hopelessly in love with him at first sight.

One day, as Narcissus sat at the edge of a lake, he happened to glance into the water and saw the reflection of an exquisite young man. Having no idea he was looking at himself, he became so entranced with his own image that he refused to eat, sleep, or move from the spot. He died fixated on the shimmering boy in the limpid water. The white flower we call narcissus was said to have bloomed below where his body lay.

It's a well-known myth—and the source of many misconceptions. People use the term *narcissistic* to describe someone with self-adoration, like Narcissus. But having known and treated many adults who had a narcissistic partner or parent, I don't believe that narcissists love themselves at all, although they may appear vain, confident, and extremely arrogant.

In reality they are deeply insecure and self-doubting. If they weren't, why would they have such an insatiable need for approval and adoration? Why would they constantly need to be the center of attention? And why would a narcissistic mother need to block her daughter's developing confidence and self-worth in order to build up her own?

Narcissistic mothers don't make us feel unloved because they love themselves too much. They make us feel unloved because they are so absorbed with making themselves seem important, blameless, and exceptional that there is little room for anyone else.

Young daughters of narcissists quickly learn that anytime the spotlight falls on them, their mothers will step in to fill it. These daughters become accustomed to being pushed aside, treated as an accessory or fading into their mother's long shadow. Their confidence and natural enthusiasm evaporate as the narcissist takes credit for their accomplishments, and blames them for her unhappiness. Her needs, ego, and comfort—not theirs, they learn—almost always come first.

The Narcissism Spectrum

The narcissism that's so destructive to daughters falls on the extreme end of the broad spectrum of behavior we label as narcissistic. To look in the mirror and say, "I look great today!" or openly appreciate and admire one of your own talents is self-protective, amplifying as it does your sense of self-worth and helping you to act in your own interests or stand up for yourself.

But a little farther up the scale, self-love edges into self-centeredness. Narcissists in this category amplify their self-appreciation with a demand that steady attention be paid to their wonderful characteristics. This behavior may be irritating, but it's not toxic. A person with moderate narcissism might be vain and noticeably self-centered, dominating conversations and not paying attention to cues that her companion or "audience" is getting restless. Yet if she's confronted or called on her actions she may apologize.

There are seldom apologies, though, when a person is severely narcissistic—a condition that mental health professionals call Narcissistic Personality Disorder (NPD).

Two of the defining characteristics of the disorder are the narcissist's grandiosity and her insatiable hunger for attention. It's normal for children to have grandiose fantasies of being powerful and adored, especially if their reality falls far short of what they yearn for, but as they gain a sense of self-sufficiency, most adults put such fantasies aside. A mother with NPD, however, has never evolved beyond these early yearnings—she clings to them because they're defenses that mask her deep feelings of inadequacy. She's excessively dependent on other people's opinions for her sense of identity and self-esteem—they're her mirror—and she's driven by the need to get others' approval. So she moves through her life preoccupied with proving (or at least arguing) that she's more beautiful, more brilliant, more talented, more desirable than other women. That, in her mind, entitles her to special treatment, and she doesn't take it well, to put it mildly, when people don't agree. She's jealous, envious—and highly defensive when challenged. As you might guess, her sense of empathy is highly impaired; she has very little interest in other people and their feelings, except as they can help her inflate her sense of well-being.

Narcissism wasn't recognized as a personality disorder until 1980. Before that, it was common to play down even extreme narcissism, waving off the behavior with labels like self-centered, conceited, or egomaniacal. Now we realize that, while severe narcissists are not crazy—that is, disconnected from reality and unable to function—they have a different circuitry board than others. No one knows exactly why that is, and mental health professionals have been struggling for years to figure out what creates this personality disorder. For a time it was believed that early trauma or overindulgence led those with the disorders to create a false self, but new evidence supports the idea that the cause is primarily genetic or physiological.

What we know concretely is that people with NPD behave in ways that are highly dramatic, emotional, and sometimes bizarre. And we know that severely narcissistic mothers are dysfunctional and destructive to their daughters.

If you recognized your own mother in the descriptions above, it was probably a relief to see the truth, difficult though it is, in full sunlight. But while a description of traits is extremely useful in identifying what you're dealing with, on paper it seems almost disembodied. It doesn't come anywhere near touching your emotional turmoil and the hurt your mother has caused you. Terms like "lack of empathy" can't begin to capture the sense of emptiness you feel when you try to get some kind of consistent understanding from a severely narcissistic mother.

A Mistress of the Three *D*'s: Drama, Deflection, and Denial

Dana: Upstaged and Ignored by the Drama Queen

Dana, a bright and charming thirty-eight-year-old, told me she was exhausted from trying to balance the needs of a husband, a job in public relations, two young sons, and a severely narcissistic mother. She said that she had a good life—except for the times when she was expected to be around her mother. This caused her a great deal of anxiety, which was spilling over into her marriage and her relationship with her sons. She told me about an infuriating recent event.

DANA: "At a family dinner I announced to everyone that I was expecting my third child. My relatives—aunts and uncles and cousins and my brother—were thrilled, and they all crowded around me, laughing and hugging. Suddenly my mother got

up from the table and did a pretend faint on the floor. It was
pretty shocking, and of course almost everyone left my side
to tend to her. I couldn't imagine what had happened to her,
and my father ran to get her some water. When she got to her
feet she looked at me and said, 'How can you do this to me?
How can you worry me like this? You know you aren't that
strong. Now I'm going to have to be taking you to the doctor
all the time!' I didn't know what she was talking about! I'm
very healthy! She never took me to the doctor with my other
children. Why did she have to turn a happy evening into such a
Greek tragedy?"

Dana was outraged, but not really surprised. The fainting incident was typical of what she had experienced most of her life with
her mother, Evelyn.

DANA: "I guess she's always been a diva. I have this clear memory
from when I was little. I was maybe five years old and some of
my parents' friends had come over. I'd been taking tap dance
lessons and I was walking around in my tap shoes because
I loved the way they clicked on the wood floor. Somebody
put on some music and asked me to dance. I was a little shy,
but I got up and started to do this simple routine I'd been
practicing. Mom practically bolted out of her chair and started
doing a very elaborate dance herself. Everybody whistled and
applauded and forgot about me. I was really confused—what
harm could it do to let a little kid have her moment? Anyhow,
that's what it was like growing up. It was *always* like that
whenever I started to get some attention. I might as well have
been invisible when she was around."

Dana told me that growing up, she felt as though anything she
did, and anything that happened to her, was just an opening for her

mother to snatch some attention. When Dana sprained her arm in fourth grade, her mother scarcely comforted her before launching into stories of her own skiing injuries—which were "far worse than this." To Dana's deep chagrin, her mother showed up at her high school graduation wearing a revealing, over-the-top dress that "made everyone stare." And even though Dana was an adult now, the drama queen behavior hadn't abated. Evelyn made almost everything involving her daughter about herself. That's what narcissists do.

Addicted to Adoration

It's not uncommon for the narcissistic mother to deflate almost physically when she's not the center of attention. Adoration is her drug, allowing her to maintain her sense of self-importance, and she's lost without it. There is an old Lon Chaney movie called *The Mummy's Tomb*. The creature at the center of the story needs the leaves of a special plant called tana to stay alive, and the monster wreaks havoc in his quest to get it. The narcissist demands her tana leaves, adoration, to survive emotionally, and she'll go to great lengths to ensure that she has a steady supply.

Because she has such a flimsy sense of self, and seems to lack a core sense of worth that would allow her to feel good about herself even if no one noticed or praised her, all is not right in her world unless she is being fussed over. There's a certain pathos in this. It's almost as though she fears she'll disappear if the people in her life look away—so she demands that they don't. Attention must be paid—and whether she reaches for subtle one-upmanship or grand performances, she is practiced at using drama to get it.

But only positive attention soothes her. Criticism, or even disagreement of any sort, triggers inner turmoil, which is so uncomfortable for her that a pair of disorienting defenses kick in immediately to make you regret ever confronting her about her behavior. First, she'll deflect any discomfort from herself onto you, so the focus always stays on what she describes as your deficiencies as opposed to her own shortcomings. That generally works well enough to ward off complaints and direct discussions. But if pressed, she may simply deny your version of what happened. Those are the narcissist's Three *D*'s: Drama, Deflection, and Denial, a crazy-making, guilt-inducing combination that guarantees that it will be extremely difficult to express your differences—or stand up for yourself.

Dana had lived with her mother's drama for so long that as an adult, she had largely come to the conclusion that complaining was futile. "I was floored by that fainting episode," Dana told me, "but I was going to let it go. It was just so typical it wasn't worth getting upset about."

Her husband, Chad, though, pushed her to make a rare attempt to protest. And what happened next is a textbook case of deflection.

DANA: "Chad saw how I was avoiding talking to Mom about what happened, and he said, 'Look, I think it's time you talked to your mother. She's been pulling this stuff a long time.' I couldn't argue with that. So I made myself go over and see her. I was pretty anxious because I've tried bringing up what she does, and I always end up feeling frustrated, even worse than before.

"I said, 'Mom, I really need to talk to you about something,' and right away I could see her tense up, but I kept going. I was proud of myself for that. I said, 'It's really hurtful and embarrassing for me that you get so dramatic when I'm saying something important about my life. That fainting thing you pulled at the dinner last week was pretty shocking. I'm not having another baby to punish you or hurt you in any way, so why did you have to act like I was doing something awful to you?'

"'I don't know why you had to get pregnant again,' she told me. 'You know how I worry about you.'

"I said, 'Mom, my getting pregnant had nothing to do with you, and that scene at our dinner was a fiasco. You always have to have the spotlight, and it seems like you can't stand it when I get any attention.'

"The next thing I knew, she did what she always does when I try to point out anything to her about herself. She took her thumb and index finger and started to rub the bridge of her nose, like she had a headache. Then she puts her head down and goes, 'Honey, this is so difficult for me. You make it sound as though I'm the worst mother in the world. I really can't handle your anger right now.'"

Dana's mother expertly shifted any discomfort she felt to her daughter, never responding at all except to say, in words and gestures well chosen for their dramatic, guilt-inducing effect, "Look how much you've hurt me."

THE DEFLECTOR SHIELD DEFENSE

Deflection is a powerful defense for a mother with severe narcissism. She uses it to keep you at a distance so she doesn't have to consider, or even acknowledge, your feelings and the possibility that she may be in the wrong.

She can't afford to let anyone challenge her image of herself as perfect. Invincible. Above reproach. Just as the Great and Powerful Oz is a front for the flawed and all-too-human figure behind the curtain, the image the severely narcissistic mother projects hides her deeply insecure core. She protects that shaky inner architecture by fending off anything that would force her to examine or question herself. It's unthinkable for her to admit there are any cracks in her facade, perhaps because on some level she knows that if she did, the whole house of cards would collapse. A healthier person, confronted with a disagreement or an unflattering image of her own behavior, might react with curiosity or doubt or sadness. She'd most likely allow the possibility of another point of view. But any time you disagree with the severe narcissist, or criticize her, her raw nerve endings tell her only one thing: She's been attacked.

Evelyn was certainly no screamer as she rubbed her nose and put her head down on the table in response to her daughter's complaints. But behind the passive front, there was plenty of heat and aggression. With body language that suggested "You've hurt me so much I can't hold my head up" and exaggerations like "You make it sound like I'm the worst mother in the world," she took the offensive and shifted the blame to Dana.

DANA: "I didn't see what she was doing. She never screamed at me or even seemed to be mad. But I can definitely see how critical and angry she was. I felt it in my body. My neck and face got hot, and my stomach clenched. All I did was try to stand up for myself. She's so good at making me feel like *I'm* the one who's out of line."

Lying, Gaslighting, and Denial

A severely narcissistic mother's anger, criticism, and thoughtless dismissal of her daughter's feelings are painful and destructive. And every daughter clings to the belief that if only her mother could *see* that behavior and its effects, she'd stop. Daughters try again and again to hold up a mirror, hoping that this time, things will be different. But severe narcissists stay true to form, responding to any confrontation with drama followed by deflection and a focus on *your* shortcomings. When that doesn't produce the desired results, they turn to what may be their most frustrating and infuriating tool: denial. Confrontation makes them feel cornered, and when that happens, they can't and won't validate your experience or acknowledge their part in it. Rather, they rewrite reality and tell you that what you saw you didn't see, what you experienced didn't happen, and what you call real is actually a figment of your imagination.

It's extremely disorienting, as Dana saw. Her mother didn't stop with conveniently developing a headache to silence her daughter's complaint about her drama queen behavior. She threw denial into the mix:

DANA: "It got worse, though. She got up from her chair and was heading for her bedroom when she looked at me and said in this really calm voice, 'You know, dear, I just don't understand how you can say I fainted. I got excited and sat down. Don't I have a right to do that? Your hormones must be making your memory fuzzy. You'd better go now. I need to lie down.'

"At that point I felt so confused and guilty that I just slunk out."

A severe narcissist is highly unlikely to admit being wrong, no matter how egregious her behavior, and she'll say whatever she feels she must to portray herself as being in the right. She'll lie about what she has promised, lie about behavior that you've witnessed, and lie about what other people have said and done. Often, as Dana saw, that involves not just lying, but also turning the tables and calling *you* a liar. She may throw you totally off balance by denying your very reality with lines like:

- That never happened.
- I never said that.
- Are you sure you didn't dream this?
- You have a vivid imagination.

Then she'll step up the attack with criticism like:

- You're so unforgiving.
- You're so overly sensitive.
- I was only kidding. What happened to your sense of humor?
- You always take me the wrong way.

As she challenges your memory and your ability to think rationally, she undermines your perceptions of reality, leaving you confused and wondering if she could be right. You may even begin to believe her lies about you.

In the classic movie *Gaslight*, a husband tries to convince his wife she's crazy by, among other things, hiding her possessions and telling her she lost them, or making small changes around the house and denying it when she points them out. When she says, "It's getting darker in here. You've turned down the gaslights," he says, "Nobody's touched the lights. They're bright as ever. You're probably not feeling well." Gaslighting is a common tool of severe narcissists. When it serves them, they'll insist that night is day and black is white. And the anger, pain, or upsetting behavior you complained to your mother about? You must have dreamed it.

Sharon: Stung by Narcissistic Rage

Severe narcissists can take on much bolder hues than Evelyn's passive-aggressive expression of displeasure. When life has disappointed them or jangled their sense of entitlement, some mothers not only make their daughters scapegoats but also lash out at them with rage.

Sharon, a single, forty-year-old doctor's receptionist, came to see me for help dealing with anxiety. She had a master's degree in business, but she didn't seem able to get a job commensurate with her education. She told me, "My panic attacks have flared up again."

I asked her if she had any clue about what might have triggered them.

SHARON: "Well, for one thing, I just had a terrible experience with my mother. It's an old story. . . . About two weeks ago, I had lunch with her. She and my father have been separated for about six months, and she was planning to write him a letter because she wanted to get back with him, and I told her I didn't think it was a good idea—they always seemed so miserable together.

"She absolutely screamed at me. She said: 'How dare you try and keep your father and me apart? You're cruel and immature—I couldn't be more ashamed of you. No daughter has ever been so cruel to her own mother.' By the time she was done I felt like the lowest person on earth."

Sharon had been bombarded with the full force of narcissistic rage. By not supporting her mother's reconciliation attempt, she had unknowingly tapped into the bottomless pit of her mother's inability to tolerate criticism, opposition, or defeat. Like so many daughters of severely narcissistic mothers, Sharon was made to take the blame for her mother's unhappiness. The message was clear: "Of course I'm unhappy—who wouldn't be with such a cruel and heartless daughter."

Yelling, screaming, and insults to your worth are common responses to even neutral comments that disagree with the enraged narcissist's point of view. You're judged as good or bad depending on whether you totally support her. And she may attack with all the fury of a wounded animal, with no thought to the effect her words have on you.

"You're No Good"

Criticism flows from seriously narcissistic mothers anytime they feel insecure, disappointed, or deflated. Like all insecure people, they build themselves up by tearing you down. If you're enjoying yourself, you must be neglecting something important, or getting in trouble. Your eyes are too small. Your nose is too big. You're too fat, too thin. Your legs are too heavy, or they look like toothpicks. They may flatter you by spinning grandiose fantasies around you, but when you fall short of their ideals—that is, when the fantasies are revealed to be just that—they criticize even more.

SHARON: "Mom got this idea when I was around eight years old that I should be a model. I was a pretty ordinary-looking kid and I knew it, but she had this idea that *her* daughter should be enough for any modeling agency. I was just along for the ride—I'd never wanted to do anything like that. Through a friend, she got me an appointment with an agency, and one of the associates spent a little time with me and then said, 'Thanks for coming in. We'll let you know.'

"A couple of weeks went by and we didn't hear anything. Of course, that didn't sit well with Mom. She called and called, and they finally told her, 'Sorry, we don't need anyone right now.' She went ballistic—and all of a sudden, it was my fault that I wasn't pretty enough! She started saying things like, 'Maybe it's that moon face of yours. Maybe it's those squinty little eyes.' I can still hear her saying that, and it's been so many years. . . . I remember how I practiced in front of the mirror to keep my eyes open wide when I smiled."

Sharon, like every daughter of a severely narcissistic mother, couldn't possibly meet her mother's expectations, something her mother never let her forget.

SHARON: "I know that her mother was horrible to her and that
was the excuse for years of humiliating me. She obviously
was so disappointed in me in almost all ways. She used every
opportunity to pick on me when I got older. I just couldn't
make her happy. I got an award in math one year, and all she
could tell people was that she'd done all my homework and I
never would've made it without her help. She would say 'Good
job' to me once in a while, but I could tell she didn't believe it.
She thought she was better than I'd ever be. I could see that,
and I couldn't figure out how to make her proud of me."

As she repeatedly makes herself feel powerful with criticism
that makes you feel bad, damaged, and small, the severely narcissistic mother is teaching you to aim low and keep your head
down. You become afraid to try, and expect to be shot down if
you do.

Sharon was very bright, and she'd worked hard to get an MBA,
but her mother, who was a bookkeeper, did everything she could
to discourage her, saying, "I don't think you're cut out to be a
businesswoman." Sharon held her self-doubts at bay all the way
through her degree program, but she couldn't bring herself to
take the next step and go for a job in the field that interested her
the most: banking.

SHARON: "It was a big deal for me to go for an advanced degree,
and I was proud of myself for doing it. But I was so panicky
about screwing up interviews at large firms that I only applied
to a couple of small places. I got two rejections, and that was
it. I really don't need that kind of stress and scrutiny. I couldn't
handle anxiety, so I wound up getting a job at a bookstore for
a while. I'd rather be doing what I'm doing now than deal with
that kind of rejection. I proved to myself what I could do by
getting the degree."

Sharon's mother had done such a thorough job of destroying her confidence that Sharon's natural interview nerves quickly spiraled into panic, and she persuaded herself that it was a sign she wasn't meant to advance. Her mother's constant theme—*you're not good enough*—replayed and escalated in her head until the only way to escape it was to shut down. So that's what she did.

Not for the first time, Sharon surrendered to the overriding sense of worthlessness that her mother had instilled in her, and let it shape her life.

When the "Bad Mother" Was Once a Good One

The less secure a severely narcissistic mother feels, the more extreme her drama, anger, and attempts to feel superior are likely to be. But there are times—often when she's gotten what she wants, when she's feeling confident, or when she doesn't sense an imminent challenge from you—that her behavior calms. She doesn't need the Three *D*'s, and she doesn't need to criticize.

During those stretches, she seems like a much different person—kinder, more supportive. Some daughters rarely see their narcissistic mother's good side. But some are haunted by the contrast between their "good mother" and their "bad mother" because they may have had long stretches of positive mothering, most likely when they were young. It's a common pattern: A narcissistic mother with relatively few stresses in her life and loads of adulation from her young daughter envelops the girl in her world, embracing the role of teacher and idol. But as her daughter gets older, the mother begins to see her as a rival, setting off a pattern of criticism, competition, and jealousy that continues through adulthood. When triggered by her daughter's emerging womanhood, the mother's insecurities about being overtaken only occasionally recede, and the habitual behaviors we've seen from the narcissists in this chapter become commonplace.

Daughters are tormented by memories of the "good mother" because once she's no longer a regular presence, it's hard to turn their mother's intermittent affection into something lasting, or recapture the closeness that once flowed so freely between them. But they twist their lives into pretzels trying.

Jan: Once Her Daughter, Now Her Rival

Jan, a thirty-three-year-old actress, supports herself with commercials and sporadic acting jobs, along with a small inheritance from her father. She is a very pretty young woman with ash blond hair, but I couldn't help noticing the dark circles under her large green eyes. She fidgeted with her bracelet as she sat across from me. After getting some background information, I asked Jan how I could help her.

JAN: "I'm a mess. I just got my first big break, a second lead on a series, but since I found out I've been so anxious, I've been eating to calm down, and I've gained seven pounds. My fingernails are gone. I can't sleep. The director said to me, 'What the hell is going on?' He told me I've got to knock off some weight. My friend Anna says I'm sabotaging myself. I have to get it together."

Clearly there *was* some self-sabotage going on, and to get a fix on it, I asked Jan if she could give voice to the anxiety she was feeling by focusing on the fears and thoughts that were keeping her on edge. What did they sound like?

She thought for a moment.

JAN: "It's like: 'Who do you think you are? You're not that pretty, you can't get into any of your clothes and you're a screwup. You're going to blow this job.'"

That kind of critical inner commentary doesn't spring full blown as the voice of truth in a woman's head, and when I asked Jan if someone close to her regularly doled out criticism, it didn't take long for her to come up with an answer.

JAN: "Well. . . . My mom's not the most supportive person in the world. I invited her to watch one of our rehearsals—I thought she'd get a kick out of that. When it was over, I asked her how she thought it went, and she said it looked like a good show. But then she looks at me and goes, 'Look, honey, I don't want to hurt your feelings, but you're no Meryl Streep.' It's the weirdest thing, because she says stuff like that a lot now, but she was so great when I was a kid. In fact, she's the one who encouraged me to become an actress. When I was young, like seven or eight, she used to take me to see plays, and not just little-kid stuff, and that's when I fell in love with acting. Those were my special days. I was so happy that my mom wanted to share what she loved with me—she'd done a little acting when she was young, and I wanted to be just like her. I idolized her. But then she changed. When I got a little older . . . it's like I lost her."

Many clients have told me of having wonderful times with their mothers when they were little, days full of hugs and laughter. And they've puzzled over how dramatically that ended when they reached adolescence. It's a crushing turnabout: You had a mother for a while, but suddenly you don't—and you wonder what the heck you did to lose her. Actually, it's simple: You stopped being an awkward, flat-chested girl and became a threat to her as a woman.

As we talked, Jan found that she could trace the change in her relationship with her mother back to high school.

JAN: "Mom started trying to be friends with my friends and my first boyfriends, and not in a mom kind of way. I noticed how she would put on lipstick before they came over, and hang out in the kitchen with us. She would act as if they were *her* friends and try to buddy up with them. And she would make snide little jokes about me, as if they were her pals and she felt sorry for them for having to be with me. When I got older, I really thought about not having my dates pick me up at home because my mother was so overtly seductive with them. She would wear revealing blouses and stand way too close to them, reeking of perfume. Once, when we were in the kitchen fixing coffee for one of my dates, she whispered, 'I could tell he would really rather be going out with me.'"

Suddenly the roles and boundaries between mother and daughter were blurred and bewildering. The competitive mother had gotten into the arena and put on the boxing gloves. Jan told me that the rivalry only got more pointed as she got older.

JAN: "I remember once when I was about seventeen—I know now that I was pretty and smart, but I was also very insecure. A boy I was crazy about had just broken up with me, and I was devastated. We were on one of those awful family vacations at some dude ranch—and I couldn't ride worth a damn. My parents and sister wanted to go out on a trail, and I went along, too. I didn't want to look like a killjoy. I was miserable and bouncing all over the place. When we got back, I sat on the porch of our bungalow feeling really rotten. My mother came over and sat by me on the steps. She got an almost kindly look on her face and I thought, 'She knows

how much I'm hurting—maybe she's actually going to say
something comforting to me.' But after a minute she sighed
and said, 'You know, dear, let's face it. You'll never be the
athlete I am. You'll never be the rider I am, and you'll never
be the woman I am.'"

Where could a remark like that come from? Jan's mother, Pam,
as I learned, was dissatisfied with her marriage, and her early ambition to be an actress had ended in frustration. So she seized the
opportunity to zero in on Jan's vulnerability. That way she could
momentarily feel superior and assuage her own insecurities.

For Jan, as for all daughters who find they've activated their
mother's competitive side when they need to be soothed and loved,
the experience was shattering.

JAN: "I was so hurt and bewildered. I kept asking myself, 'What
   did I say? What did I do? What's wrong with me? Why
   doesn't she love me anymore?' And those words of hers. I can
   still hear them. . . . I just wanted to curl up in a little ball and
   disappear."

Jan kept pursuing her acting, first in school plays, then in community theater and small professional jobs in television, sure her
mother would be elated and that she'd win her back. But the response was almost always the same: criticisms and slights instead
of encouragement. The woman who had once been her biggest fan
now said things like, "I'd love to help you with your lines, honey,
but I get so impatient with your stumbling. I always thought you
had my good memory, but I guess not. . . ." The message was loud
and clear: Anything you can do, I can do better.

JAN: "The implication was that I could never measure up, and it
   really hurt, because I thought this was something we could

share. I was so confused. She created this huge desire in me to be an actress, and then when I actually went for it, it was like she didn't like it because I was challenging her or something. It's pretty much been like that ever since."

WHAT'S BENEATH HER NEED TO COMPETE: EMPTINESS

Reasonably healthy, fulfilled women don't have the need to compete for their adolescent daughters' boyfriends or squash their fledgling attempts to try out their passions and take risky first steps toward becoming the kinds of women they want to be. They see their girls in the most vulnerable and self-conscious time of their lives, remember their own stumbles, and try to ease the way.

Narcissistic mothers like Jan's can't connect with that sense of compassion, not only because of their insecurities but also because at their core they have a terrible sense of deprivation, an insatiable hunger that makes them believe there will never be enough for them, and that anyone else's gain—even their own young daughters'—will keep them from adequately filling the hole inside. In some ways, they're like the "hungry ghosts" described in Asian culture: creatures with enormous stomachs that ache to be filled, but minuscule mouths and narrow throats that leave them feeling perpetually empty. That's a good picture of the insatiable hunger of these mothers, who greedily grab all they can, from anyone they sense is cutting into their supply of men, money, respect, affection. Whenever they sense you as a competitor, you become a constant spur to their longing.

Where does this distorted sense of "not enough" come from? The likeliest answer is competition and a sense of scarcity in the mother's own background. She may have had a competitive mother herself, and grown up with the disorienting sense that she couldn't get or be what she wanted without somehow depriving her mother, or fighting her off. Or she may have come from a family in which there was intense sibling rivalry, a setting in which she had to compete with her brothers and sisters, cousins, or extended family members for love and goodies.

This emptiness and fear of deprivation are often well hidden under a seemingly confident exterior as she explains her sometimes desperate grabs for what she wants with the narcissist's typical rationale that "I deserve it because I'm superior." It's a claim that can't stand scrutiny—and it would be more accurately stated as "I deserve it because I *need to feel* superior"—but this mother isn't likely to spend much time scrutinizing her own motives or questioning her assumptions.

You Absorb Her Ambivalence About Your Success

A daughter like Jan reaches adulthood having been steeped in the ambivalence and envy that her mother offers her in place of encouragement and support. She was lucky enough to receive praise when she was young, but she becomes mistrustful of it when she's older, because she's seen how often it's followed by put-downs. And she internalizes her mother's puzzling "go for it—but don't get your hopes up; you're not good enough" attitude.

JAN: "I'll never forget getting my first commercial. I was so excited I was telling everyone about it. I made the mistake of inviting my mom to dinner to tell her the good news, and as soon as I did she says, 'That sounds wonderful, dear, but don't expect too much to come of this—you're just not that photogenic.' I don't want you to get the wrong idea about Mom, though—she's got her good side, too. As soon as she tells me how bad I look on camera and lets that sink in, she does a complete one-eighty and says, 'But come on, we'll fix you up.' She pulls out her car keys and goes, 'I saw a sweater at Nordstrom that will bring out the green in your eyes. That'll get their attention.' And she buys me these amazing clothes. I don't know what it is with her."

Despite her cuts and snipes, the narcissistic mother sometimes *does* seem to want you to get what you're after. Her gifts may come with barbs ("Let's fix you up"), but she occasionally offers them, perhaps because she wants to return to the teacher/idol role she enjoyed when you were little. And at least momentarily, she enjoys the reflected glory of your success. After all, she's your mother, and she can take some credit—even most!—for your accomplishments. Your success is also often a screen onto which she can project her fantasies about being young, desirable, capable, and talented.

"It sounds like you're getting a lot of conflicting messages from your mother," I told her. "It's like: 'I'll help you go for it so I can live vicariously through what you're doing, but please fail or let me overshadow you so I can feel better about myself.'"

JAN: "Oh my God, that's exactly like my mom. I can see she's yearning to be doing what I'm doing, and she wants to help me make it. She thinks it's glamorous and exciting. But at the same time, she *doesn't* want me to do well or ever have a special moment of my own. I think it makes her feel like a loser. It's bizarre. She puts me down, but she envies me."

For Jan, the intermittent generosity from her mother helped drive the self-defeating hesitation she brought to her work. If she did well at an audition, she'd please the mother who bought her expensive clothes—the one who seemed so much like the good mother who'd encouraged her when she was young. But any real success would trigger her mother's jealousy, and all its repercussions. An adult daughter who longs for renewed closeness with her narcissistic mother frequently considers such alternatives and stalls at the threshold, clueless as to why she's procrastinating on a high-profile project or putting on weight on the eve of an important appearance. The process isn't rational, and for the most part, it's not conscious. What you experience is the push-pull sense of wanting to succeed but being held back by some mysterious force, which is often a deep sense of guilt. Your mother has taught you that you can't, and shouldn't, go for what you want. You've learned her most important lesson: *Don't outshine your mother.*

She Fans the Envy in Your Family, and in You

Another all too common effect of growing up surrounded by so much jealousy is feeling jealous yourself. Often daughters absorb their mothers' bottomless hunger for what other people have, and take up the envy baton that's been quietly passed to them:

JAN: "I was boy crazy from the time I was about fourteen. I turned to boys to get out of the house—I kind of found myself through my relationships. But if there was a time when I didn't have a boyfriend and some of my girlfriends did, I would feel angry and depressed. It was like how dare they have what I need so much. It can still happen if I don't have a guy and one of my friends does."

Jan told me that her mother actively fans those sparks of jealousy even now by comparing her with other people.

JAN: "Mom likes to send me clippings from the newspaper or magazines about other women's marriages and successes. Or she calls me up and says things like, 'Did you hear about your cousin Amy? I heard her new boyfriend is taking her to the south of France for three weeks. . . .' I sure didn't want to have that conversation, so I said, 'That's nice for her,' hoping that would end it. And my mother's reply was, 'It sure is. . . . Why can't *you* find someone like that?' She made me feel terrible, and it was impossible for me to feel anything but resentful of my cousin's good fortune. I hated feeling that way."

There's no need for her to say it directly. The message you take in is all too clear—you've lost a race you didn't even know you were in. You're not as pretty or as sexy as your cousin. What's wrong with you?

If you have brothers and sisters, your competitive mother may encourage lifelong rivalries among you that give her the superior sense of being in control of the outcome, and therefore the winner, of what may be a charged contest for her approval.

On her whim, one child may be deemed the golden one who can do no wrong, while another becomes the family's scapegoat. If you've frequently been cast in the scapegoat role, you may suddenly find yourself in favor, close to her for a time—just as you were when you were young. But if something—your spark, your smile, your solo in the choir—threatens her, you'll quickly find another sibling in your place.

As you and your siblings grow up, she often keeps you engaged in the family loyalty battle by dispensing and withholding favors in the highly charged arena of money, gifts, and inheritances. These battles may well be a window into the roots of your mother's sense of deprivation; it's quite possible that she's reenacting old patterns between her own sisters and brothers when she manipulates you and your siblings. But this time, while her own kids start to be jealous of one another, she can remain above the fray. This time, *she* wins.

You'll Never Be Able to Please Her

Despite all this, many adult daughters of severely narcissistic women hold fast to the hope that they can repair their relationships with their mothers and that their mothers will somehow become more loving.

What you *want* to believe is that she's got your well-being at heart. And the intensity of that desire can take you by surprise.

JAN: "I was over at my mom's the other day, and after lunch, she told me she'd found an old album in the back of the linen closet. She had put it on the coffee table and we started to go through it. It was full of old photos of me as a little girl, and some shots from a trip we took to New York when I was little. I hadn't seen those photos in years. We sat there looking at them for the longest time, and they brought so many memories back. I can't believe it, but I miss that mom so much. I just wish I could make her happy."

I'm sad to say that that's highly unlikely. Narcissistic mothers are close to impossible to please.

Daughters resist accepting this. They keep hoping for the perfect words, the perfect gesture, that will let them hear the words "Thank you" and "I love you" from mothers who so rarely express real affection and gratitude. Dana, the daughter of the drama queen mother, whom you met at the beginning of this chapter, told me this poignant story:

DANA: "I decided to throw a birthday party for Mom's sixty-fifth. I was going to make it very special—have it catered and decorate the house with balloons—I thought she would love that, she would be the center of attention, and that would really please her. I invited several family members and some of her friends.

"I had spent several days looking for just the right present. I knew she liked Asian antiques, and I finally found an exquisite old piece of Chinese sculpture. I had to dig into savings for it, but I figured, 'What the hell.' As soon as she opened it, it was obvious from the look on her face she didn't like it and she made no attempt to hide her reaction. After everyone left, I felt pretty let down. The next morning she calls up, and of course I thought she was going to thank me—it really was a lovely party—but instead the first thing she says is—not even 'Hello, how are you?'—it's, 'Why did you have to let everyone know how old I am? Some of the people there didn't know. Did you deliberately set me up to be humiliated?'

"I just wanted to cry. Nothing I do is ever enough."

Even the most well-intentioned act or statement can be distorted through the narcissist's self-referential lens and her insatiable need to look good. If she perceives that something was meant to embarrass or diminish her in any way, you're likely to find yourself facing her suspicious accusations. The relationship between narcissism and paranoia hasn't been fully explored. But when the narcissist takes one of your benevolent gestures as a deliberate attempt to embarrass her, you can feel the connection.

What You See Is What You Get

Mothers with NPD sometimes raise your hopes by agreeing to go with you into therapy, but they do not respond well to the process. They lack two crucial elements for change—self-awareness and the ability to be introspective—which makes counseling all but a charade. As long as they can blame everyone else for not filling their insatiable demand for attention and adulation, they can successfully avoid responsibility for their own damaging behavior. They're good at that, and because they rely on it to feel better, they have no reason to change.

These mothers are in the grip of a deeply ingrained personality disorder. And that behavior is not just situational—it is at their core.

Please don't forget, as we explore this difficult territory, that your own core is very different from your mother's. The harmful behaviors you've learned from her and the pain you've carried with you for so long are not a permanent legacy. As I will remind you throughout this book, despite what she's told you, you are the healthy one. You can change.

Chapter 3

The Overly Enmeshed Mother

♥

"You are my whole life."

You've probably heard of the well-known humanitarian organization Doctors Without Borders. The far less noble group of women you'll meet in this chapter are Mothers Without Borders. The enmeshed mother looks to her daughter to fulfill her need for companionship, give her a meaningful identity, and provide vicarious excitement. You are her everything.

At times the closeness the overly enmeshed mother offers seems to be just what every daughter of any age craves. There's a certain warmth between the two of you, and there can be genuine appreciation of you and your accomplishments. But her definition of "closeness," you discover even when you're quite young, can be suffocating, invasive, and unilateral—she insists on it whether it feels good to you or not. The ultimate mother who can't let go, she presses herself on you, co-opts your plans, and plants herself in the center of your world, believing that she's behaving lovingly. As you grow older and try to shape your own agenda, letting her know that you have needs and wishes of your own, particularly ones that exclude her, she rarely releases her grip without a fight.

Like all unloving mothers, she puts herself first. Even if you have a full life of your own, she wants you to stay her little girl, joined to her at the hip. She holds out promise and praise that disappear when you prove to have a mind of your own. And she tries to mold you by making you feel guilty if you don't go along with her wishes and needs.

Trish: How Bonding Turns to Bondage

Trish, a twenty-six-year-old teacher's aide, called me because there had been so much tension in her family after the birth of her first child. I asked her what she thought was causing the strain.

TRISH: "I had been thinking for some time that my mom needed to give me a little space, and Doug, my husband, had been complaining about the way she always wanted to do things with us, no matter what we had planned. I'm used to her—we've always been kind of inseparable, for better or worse. But after what happened when our baby Lily was born. . . . I hate to admit it, but he's right—she's out of control."

I asked Trish to give me an example.

TRISH: "I was in the delivery room. I only wanted Doug with me, and he told my parents they had to stay in the waiting area. My mother got very upset and said she belonged in there with me. Doug told her very politely but firmly that this wasn't going to happen. There was a bell on the door to the delivery room, and to my horror my mother kept pushing it every two minutes. When a nurse opened the door she demanded to be let in. The nurse said I didn't want anyone in there and my mother started crying. 'I need to be in there with my girl,' she kept saying. 'My little girl needs me.' The nurse closed the door, but Mom kept ringing the bell. My husband finally had to go and actually physically restrain her. She just couldn't stand being away from me, which sounds like a good thing. But I didn't want her there. I just wanted Doug. He's really upset about what happened, my mom's not speaking to me, and I feel guilty as hell."

The pressure, the tension, and the guilt were all familiar to Trish. Her mother, Janice, had been in nursing school when she got pregnant with Trish, and she dropped out to raise her. "Mom gave up everything for me," said Trish, repeating the familiar family story. Disappointed with her marriage and with no career, Janice felt a deep void inside herself, Trish told me. But she still had her daughter. Trish became her companion, confidante, and reason for being.

TRISH: "I remember when I was eight and we were riding the subway. We'd just come from a movie. She put her arm around me and said, 'You're absolutely my best friend. You're so smart, such wonderful company—I'm so unhappy with your father.' I was so proud, but part of me was really uncomfortable. When you're eight, you don't want to be Mom's best friend. You want her and your father to be close, and you want her to have her own friends. You just want to be her little girl."

Trish told me that her mother's marriage had always been troubled. Janice married the young man who got her pregnant, and they were never a good match. He started staying out late and having affairs soon after they were married, creating an atmosphere in which Janice needed to find nurturing somewhere else. And she turned to Trish for almost all of it. She could escape into the company of a little girl whose complete and uncritical affection was as close to unconditional love as she could get.

So Janice enveloped her young daughter with what looked and felt like adoration. Here was her mother saying she'd rather be with her than anyone else—how could that be so bad? But even at eight Trish knew something was not right.

A mother like Janice is devoted, not neglectful, when her daughter is young, though she may hover, determined to buffer her baby (and that's how she'll often refer to her child, whatever her age) from disappointments and difficulties. She'll fight to get her child a good grade, or an invitation to a birthday party, or the status item everyone wants. None of this seems unloving. But it can prove to be exceedingly so as soon as a daughter tries to break away, explore, and express her own desires. That's when so much of what the mother believes is closeness, love, and bonding reveals itself to be an elaborate form of bondage.

In a healthy relationship, the bond between a mother and her daughter is meant to be a flexible, malleable connection that can withstand distance, conflict, and differences—differences of opinion, feelings, needs, desires. Ideally when a child first tests that bond by trying out the word "no!" around the "terrible twos," she discovers that even when she asserts herself and defies her mother, the love between them doesn't disappear. It's safe to be her own person, and she can trust the bond with her mother will be there.

As the child grows, she takes bigger steps into the world on her own, falls, and makes mistakes. And, if she's lucky, her mother is the safe harbor she can return to, even after doing something foolish or rebellious. This is especially true during the teen years, when a daughter is figuring out who she is, testing limits, learning what those alien creatures called boys are like, and deciding what kind of woman she'd like to become. A loving mother-daughter relationship may be frayed, rocky, and tumultuous at times, but there's a steady undercurrent of acceptance, which helps give daughters the courage to grow, evolve, and become separate individuals.

That's not what enmeshed mothers have in mind. Many of them have made motherhood not only their entire definition of themselves and their value, but also a way to soothe their own very common fear of abandonment. Some may have partners, careers, and friends of their own, but what eclipses all of that is their role as the mother of a dependent child who needs them, and even feels like the missing piece that completes them. The "closeness" they want is so all-encompassing that a daughter, as the familiar phrase puts it so well, often doesn't know where she stops and her mother begins.

Enmeshers place the burden of their happiness on you, and instead of teaching you to build a life of your own, they snap on the emotional handcuffs, and never let you go.

Separation Is Not Allowed

Overly enmeshed mothers see the very normal and necessary process of separation as a loss and a betrayal, and they work hard to pull you back in anytime you try to grow up, pull away, or leave.

Natural transitions, like a daughter's move from home to college, frequently trigger what feels like the empty nest syndrome on steroids. When Trish graduated from high school—and long before that—Janice had many options for a better life. She could have gone back to school and picked up her vocation or sought out marriage counseling for her husband and herself. Nothing was holding her back. But by this time, she was so used to reaching for her daughter to fill every bit of her emptiness, that's where she continued to put her energy:

TRISH: "It was so embarrassing when I went away to school. She made a federal case out of the fact that I wanted to go to a school in another state, and one of the reasons she was willing to stop fighting me on it was that her sister lived in my college town, so Mom could use visiting her sister as an excuse to check up on me. Mom had the irritating habit of just 'dropping by' to see me, and calling at all hours. I'd come in late and the phone would be ringing. It was always her, wanting the play-by-play of my most recent date. Thank God there were no cell phones then. Now Mom's cell is like her drug—she calls me constantly, she texts, she wants to Skype, especially now with the baby. It's horrible to say, but I feel like she's in my pocket, spying on me. Mom GPS—she always knows where I am."

Mothers like Trish's may constantly repeat, "I'm so glad we can share this experience" and "I'm so glad I can be there for you," but they rarely ask if their presence is welcome. They frame their neediness and the claustrophobic world they've engineered for the two of you as a "special gift" that other daughters would love to have.

And daughters learn that it's their job to keep their mothers happy by sticking close and keeping Mom at the center of their lives.

Stacy: Caught In the Strings Attached to a Mother's Gifts

As frustrating as it can be to be smothered in this way, there are times when enmeshment can feel like love, at least for the moment. Suddenly, just when you need it most, the overly enmeshed mother may offer money, resources, or experiences—and that can seem like a godsend.

But there's usually a catch.

Her gifts, some quite generous, inevitably create a sense not only of obligation to her but also a dependency that can be crippling. By keeping you from having to stand on your own two feet, she makes herself indispensable. That can be her license to move in and take over, sometimes almost literally.

Stacy, an athletic thirty-seven-year-old who was newly married to the owner of a small construction company and working in his office to help out, came to see me because her husband had given her an ultimatum about her mother Beverly's constant invasion of their lives. Stacy was confused, she said, because her mother had been a great help to them. Their construction business was strug-gling in the weak economy, and they needed stability, especially for their two children, Stacy's eight-year-old son and six-year-old daughter from a previous marriage. The last thing she needed was friction with her husband.

STACY: "He actually told me I had to make a choice between my mother and him—he said that he didn't expect to be married to two women! He said he loved me deeply and he didn't want to break up the marriage, but that she was driving him crazy and he can't stand how I almost disappear and get so timid when she's around. He told me he can't stand how angry and resentful he feels. I love both of them, but I feel like I'm between a rock and a hard place."

I asked Stacy to tell me how things got to this crisis point.

STACY: "I guess it came to a head when my mother bought the house next door to hers—she made a lot of money as a real estate broker—and offered to rent it out to us at a very low price. Well, here we are just starting out together and we're struggling because the income from Brent's company and my job is barely meeting our monthly expenses, and my mother buys this really nice house that we can rent for practically nothing. So I thought, 'Great!' She sweetened the pot by saying, 'I can help with the cooking and be there when the kids get home from school and I can save you tons of money.' It seemed like a terrific idea at the time. Plus she would be a lot less lonely with all of us so close—she and my dad finally broke up a few years ago, and my brothers have lived out of state for years, so the kids and Brent and I are the only family she has. I could tell she was restless after she retired. . . . It seemed like

such a win-win. . . . She'd be happy, we'd get a break we really
needed. Brent was pretty resistant to the idea of having her
right next door, but I pleaded so much he finally gave in."

Although it's rarely a terrific idea for two generations to live on
top of each other like that, it would have had a chance of working out, at least temporarily, if Beverly had been respectful of the
couple's privacy and need for alone time. But she did just the opposite.

STACY: "She's there all the time. We try to be polite because
she's helping out a lot, but now we're the Three Musketeers.
If we go to dinner and don't invite her, she'll give us the silent
treatment for hours. We gave her a key to the house so she
can get in while we're at work, but she'll walk in at almost any
time, day or night. I cringe anytime I hear, 'Hellooo, anybody
home? There's a great movie on TV tonight and I really want to
share it with the two of you. . . .' She won't go back to her own
place until late almost every evening. By that time, we're so
exhausted we just fall into bed. Here we are married less than
two years and our sex life has gone to hell."

Stacy was finally starting to realize that low rent and help with
meals and child care were actually costing her a great deal. Beverly
had practically moved in with her, and Brent and Stacy's marriage
was on the line.

Beverly, like all overly enmeshed mothers, was behaving as
though Stacy had no emotional needs of her own. By making herself all-important to Stacy, as she had always done, she managed to
move into her daughter's living room, and her marriage. She could
tell herself, and her daughter, that she was just doing the moth-
erly thing and helping out during a stressful time. That help was

undeniably real, but for both mother and daughter, it served as a rationale for Beverly's demand for constant contact with Stacy. The more Stacy "owed" her mother, the more guilt she felt about claiming her basic adult right to an existence of her own. And for her part, Beverly felt more entitled than ever to claim the dominant place in her daughter's life.

The Trap of "Let Me Do It for You"

The pattern wasn't new. Beverly and Stacy had long been wrapped in a tightly woven net of dependency.

STACY: "I was the problem child in my family. I struggled through school until sixth grade, when we finally found out I had a mild learning disability. Mom thought I was lazy, and she was always trying to motivate me or find some new program to help. I was her project, I guess you could say. She did a lot for me—sometimes she'd even do my homework. She was a huge help, but she always treated me like I couldn't do anything for myself, even though I was good at sports. She was really focused on what I couldn't do and on fixing me. I know I needed it, and I'm glad she was there, but I felt like I couldn't make it without her. She would always say things like, 'Are you sure you should take that drama class? You're going to have to read the script.' I felt really stupid. Finally one of my teachers suggested I get tested, and we found out I had dyslexia. It was a big relief for me. I got tutors and they put me in special ed, and it really helped. But since reading was hard for me, Mom kept treating me like I couldn't do anything for myself. She was so overprotective, I wouldn't have been surprised if she wanted to go out with my friends and me so she could read the menu for me when we ordered pizza."

It's not easy for a mother to see a child struggle, and it's natural to step in and do whatever she can to help. But in healthy relationships, independence is always the goal. For all the help her mother gave her, Stacy grew up feeling inadequate, always focused on her weaknesses rather than finding ways to develop her strengths.

Instilling in a daughter the feeling that "I can't make it on my own" holds her back and creates a large confidence void that an enmeshing mother can rush in to fill. Almost without realizing it, a mother may come to take such satisfaction in the feeling of pride and competency that comes from "rescuing" a daughter that she loses sight of her child as a three-dimensional person. As Stacy noticed, she became her mother's project, a bird with a broken wing who would always need to be carried. And the more both of them saw the relationship that way, the less Stacy could build a full, self-defined life of her own.

For Stacy, very little changed when she graduated from high school. Her mother persuaded her to live at home while taking classes at a community college, and swooped in to give her a job at her real estate office when Stacy had trouble her first semester and decided to drop out. She kept her daughter in a bubble, never letting her fail. And Stacy never had to learn to persevere, work through difficulties, and get up after a setback.

STACY: "I finally got a little momentum of my own when I got
    bored working for Mom. I had no interest in real estate, and
    driving around to open houses was not my idea of a good
    time. I got a job at a gym and started doing some work as
    a personal trainer, which I had always wanted to do. Then
    I met a guy, one of my clients, and long story short, we got
    married pretty fast. I thought I'd finally be on my own, and
    I was pretty happy about it. We had a lot in common, and it
    was great for a while. But after Tyler was born, we started
    having problems. It started when I told him he wasn't doing

enough around the house. He was taking me for granted, and he started to lose his temper a lot. Stupidly, I talked to Mom about it. Of course she took my side, and the more I talked to her, the less I could see Mark's point of view. Mom kept saying, 'Just come home. You don't have to take that.' The whole thing blew up, and I wound up taking the kids and moving back to Mom's a couple of times. She said we could stay there as long as we wanted, and she'd help me out with money. Mark was so furious the second time, he said he was done with her and me. We couldn't put things back together after that."

The marriage broke up, and Beverly had her daughter back under her roof. That was a low point for Stacy, but recently, she told me, she'd thought she'd turned a corner in her life. Her mother seemed to like Brent, her second husband, and Stacy breathed a sigh of relief when her mother came to their rescue with the offer of the house.

But nothing had worked out as she'd hoped. For one thing, Brent was understandably feeling like a wallflower in the dependency-driven mother-daughter tango. And Stacy, confronted with the reality of wanting to please the two adults in her life who meant the most to her, was paralyzed, as so many daughters of overly enmeshed mothers are. Who could she risk alienating, her mother or her husband? She felt that she was being pulled apart in a life-and-death tug-of-war.

STACY: "I realize that as an adult, most of my major decisions have been based on what would please my mother—what would make her happy. I also realize that sometimes I've put my mother before my husband—how sick is that? That's what I did in my first marriage, and now I'm right at the edge of the cliff with my second."

I told Stacy I could help her move away from the precipice, but first she'd have to want to be an evolved, assertive woman. Like so many daughters who are enmeshed with their mothers, she'd been stuck in a half-child, half-woman identity.

Lauren: Learning to Accept the Unacceptable

Lauren, a recently divorced forty-six-year-old stockbroker with two teenaged daughters, came to see me because she was having a tough time handling the divorce, the new burdens of being a single mom, and the stresses of her professional life. But she soon revealed that a lot of her stress was coming from another source: her mother's long-standing ritual of ignoring her plans and privacy.

LAUREN: "I have some anxiety during the week, but it really
    flares up on the weekends. Saturdays I do stuff with the kids
    and usually go out to dinner either with a date or girlfriends.
    It's Sunday that's the problem. . . . My mother and I have
    this tradition that we've had for a long time—even when I
    was still married. She comes over for lunch and often stays
    through dinner. My father died of cancer about eight months
    ago, and she really hasn't done much to get on with her life.
    She says she's just living for me and the girls. Mother comes
    over right on the dot of noon. I dread the weekends. I start
    feeling anxious Saturday morning, so her visit ruins the whole
    weekend for me. She sucks the air out of the room with her
    neediness."

I asked Lauren to give me an example.

LAUREN: "It was last Sunday, Mom was over, of course. I'm in a
    group that raises money for the L.A. Philharmonic, so I got an

invitation to a lovely event that the Phil was putting on. I was fixing lunch and my mom as usual was snooping around, and of course she sees the invitation on my desk. I was planning to go by myself because I thought there might be some interesting men there—I knew I was in trouble as soon as she came in the kitchen waving the invitation."

SUSAN: "Wait. What's she doing snooping in your things?"

LAUREN: "Oh—she's done that since I was a kid—I guess I'm just used to it."

Lauren told me that her mother had always insisted that there be "no secrets" between them, by which she generally meant very little privacy.

LAUREN: "I don't know why she was that way, but I heard the 'no secrets' stuff a lot. I didn't think twice about it when I was little. But it became kind of a big deal when I was in fourth grade and I had my first really close best friend, Anna. We practically lived at each other's houses, and there was one afternoon when we were being silly and giggly—I think she had gotten a note from some boy she liked—and I closed the door to my room. A few minutes later, Mom pushes it open and in a loud and affected voice like a kindergarten teacher she says, 'No closed doors, thank you very much!' She comes in and starts going through the records we were playing, the games we were fooling around with. Then she sits on my bed and acts as if she wants to join our conversation. . . . Finally I said we were going to ride our bikes, just so we could get out of there. Anna and I laughed about it, but she stopped coming over as much. I felt like such a baby with that 'no closed door' thing. That's my mom. . . . When I was a kid, she hated it if I was in the bathroom and the door wasn't open a crack so she could 'talk to me.'"

The boundaries between Lauren and her mother had been fuzzy, at best, for years, and I told Lauren that one of our first goals would be to help her get "un-used to" allowing her life to be an open book for her mother, who was still reaching uninvited—and undeterred—into her daughter's life. Lauren told me that when her mother discovered the Philharmonic invitation in her kitchen, she pushed her way right into the evening.

LAUREN: "She goes, 'This sounds like such a nice party. Since your father died I never get invited to anything this interesting. You know how I love being around cultured people. . . . Please don't shut me out.' Then she says, 'Wouldn't it be fun to go together? . . . You know, sweetheart, it'll be just us girls.' She puts her arm around me and says, 'I'm so lucky to have you.'

"She caught me off guard, and like an idiot I told her the truth and said I was planning to go alone—I should have said I had a date, but I have a hard time lying to her. . . . And by that time, I felt so guilty I couldn't say no to her—but then I never could. . . . So I took her with me and had a rotten time. She wouldn't let me out of her sight. She might as well have put a leash on me. I feel so smothered and manipulated by her. I can't live my own life. Every part of me is aching to be free, but I just can't do it. What's wrong with me, Susan?"

I told Lauren that there was nothing wrong with *her*, but that there was a lot wrong with her relationship with her mother. She had given her mother license to engulf her life and she would have to learn how to cancel the permit.

The Enmeshed Mother's Rules of Love

With "closeness" as her all-access pass, the enmeshed mother commandeers your space and time. She may give lip service to your right to privacy, but she still ignores it. Because she sees herself as your "best friend," she feels entitled to read what's on your desk, go through your drawers, join your parties, invite herself along for drinks, and even slip into your house without asking.

She may ratchet up the invasion by subtle degrees if she's widowed or divorced and understandably feeling sad, bitter, angry, humiliated or rejected. She expects you to be the one who eases her loneliness and takes up the slack in her social schedule, essentially replacing her partner.

She co-opts and distorts the language of love as she fuses her life with yours. This isn't just a matter of semantics. If you look at the behavior behind the words—the fine print that goes with her "I love you," "We're so close," and "You're my best friend"—you'll find a long list of conditions, restrictions, and rules that have little to do with love and more to do with erasing your separate identity.

To an engulfing mother, love means:

- You are my everything, and that makes you responsible for my happiness.
- You can't live without me, and I can't live without you.
- You are not allowed to have a life that doesn't involve me.
- You are not allowed to keep any secrets from me.
- You must never love anyone more than you love me.
- If you don't want what I want, it means you don't love me.
- "No" means you don't love me.

Much of the love she feels for you is desperate, clinging, and restrictive. And this is the love you know and expect. Instead of understanding that love is a free exchange of support, encouragement, acceptance, and affection, with lots of space to breathe, you've learned that you must earn love by giving other people what they want, like it or not, and taking your own needs and wants out of the equation.

An overly enmeshed mother rarely allows her relationship with you to evolve beyond one in which she dictates the terms. It's tremendously important to her that roles not shift, and that you do not outgrow your willingness to accept being swallowed up by her. "Keeping things the way they've always been" makes her feel safe, comfortable, and in charge, and she clings to rituals that reinforce her identity as your mother and her powerful place in your life.

Rituals are not, by themselves, inherently unhealthy. Some repetitive behavior can provide a warmth and familiarity that is nourishing. Traditions like having turkey on Thanksgiving, attending a house of worship on a regular basis, or planning family get-togethers on important occasions can provide a great deal of pleasure when they're done out of choice. But when they're done by rote and driven by guilt, they feel like the bars of a cage.

LAUREN: "I have to check in with Mom every night and tell her about my day. She's so disappointed and upset if I happen to miss a night because I'm busy—it's easier just to make sure I do it than go through a long explanation. It's like an obligation I can't get away from. . . . I promise myself I'm going to confront her about this and start to set some limits, but somehow my fingers dial the phone every night and I'm right back in the same rut."

It's almost impossible to say no to someone you are bonded to with this kind of ritual when the bonds are secured not only with the natural love you have for her but also with fear, obligation, and guilt. This unholy trinity is essential to the enmeshed mother, and you'll frequently hear daughters using words like "I feel so guilty if I don't do what she wants" or, as Lauren did, "It's like an obligation I can't get away from."

When you believe that love means making the other person happy at all costs, then to love means giving up the right to your own desires. And if you veer from that version of the relationship, then fear, obligation, and guilt click into action. Fear that you'll lose your mother's love and affection. A sense that you're obligated to do what it takes to make her happy—because that's your role as a daughter. Guilt about doing anything that will hurt her feelings or upset her, guilt about expressing *your* true feelings, guilt about any complaints you've had, any smothering you've resisted.

That potent mix is the Krazy Glue that keeps daughters stuck to engulfing mothers.

Enmeshment Is a Two-Way Street

Each daughter we've met in this chapter says she's angry and frustrated and yearns to escape her mother's smothering, so what holds her back? Why doesn't she say "Enough!" What is she so afraid of?

A woman may be twenty-five or thirty-five or fifty-five years old chronologically, but *emotionally* the daughters of enmeshers are much younger. In fact, there can be a dramatic disconnect between the competent, effective woman on the surface and the scared little girl inside, who is still paralyzed by the primal fear that all small children have: If I pull away from my mother, she'll stop loving me *and I can't survive if that happens.* Daughters who have repeatedly been rescued face the additional challenge of feeling insecure about their ability to keep their lives on track without their mothers' help.

Years of living with this dependency make it feel normal, and unknowingly, daughters sign a lifetime contract with their mothers that hands over much of their autonomy and big pieces of their adulthood. When the healthy part of you chafes or complains, you may even go as far as believing "I can't survive without Mom." If you encounter her disapproval or disappointment, giving in to your mother seems like the only reasonable choice.

Enmeshed mothers are masters at using guilt. They often collect injustices, lining up instances that have displeased them and citing them as reasons why you need to do more for them. And they'll do it in the nicest possible way. They'll say: "I was really counting on having lunch with you. I'm so disappointed. . . . You didn't tell me you were going to that movie, and you know I wanted to see it." They needn't raise their voices, and sometimes they don't even need to say a word—from the time they're young, daughters are well practiced at reading volumes of meaning into a mother's look or glance. Their mothers hardly need to use fear and obligation to get what they want, because most daughters would do anything to avoid the guilt that says: "Letting your mother down is the worst thing you can feel."

You may feel criminal if you try to cancel a casual date with your mother so you can get a massage at the end of a grueling week. She has programmed you to believe that putting yourself first is a crime—and trying to skip brunch, spending time with your boyfriend, or being alone with your thoughts are major felonies.

It can be hard to see enmeshment clearly when you're caught in it because it's been with you so long it's the reality you know, but with even a little distance, your adult eyes can recognize these patterns for what they are: an extremely unhealthy exchange of neediness.

The truth is that there's no growth, no safety in this stifling symbiosis. And there are no psychological adults here, just clinging, frightened children.

Adults Have Options and Freedom

If you had an enmeshing mother, you may carry with you a great fear of abandonment or separation. You may be overly clingy with partners or your own children. You may hold yourself back because you lack confidence in your own abilities and resilience. And you may know precisely how to make your mother happy but struggle to satisfy your own soul.

Chapter 4

The Control Freak Mother

ⵛ

"Because I said so."

- If you marry that man, you're no longer part of this family.
- If you take that ridiculous job and move, you won't see another dime from me.
- Don't expect any help from me again if you don't send the kids to Catholic school.

These are just a few examples of the stinging, domineering words of overt control. There is nothing subtle about it. There is no "I love you" manipulation, those directives masquerading as affection we saw with the overly enmeshed mother. Overt control is authoritarian and often irrational. It's full of belittling and bullying. There are very direct orders, and warnings that disobedience is grounds for serious consequences.

Control is appropriate when a child is small. Children are impulsive, with no life experience, and they need protection. There are plenty of hot stoves and traffic-filled streets they haven't learned to respect yet, and rules and a mother's firm no are a valuable part of teaching and guidance. Control at that stage not only makes a child *feel* safe, it provides actual safety. But an important part of the parenting process is gradually stepping back to let a little girl learn for herself, and when a mother's control precludes her child from doing that, it ceases to be helpful and loving.

The control freak mother keeps a heavy hand clamped down on a daughter for as long as she can—often deep into adulthood— with toxic effects. Just like the enmesher, she habitually returns to behavior that keeps you dependent, then takes advantage of your dependence. And all the while, she often insists that "it's for your own good." But the unhappy truth is that pushing you around satis- fies her, and gives her a feeling of power that is often missing in the rest of her life. For the control freak mother, keeping you locked in that power imbalance is key to her happiness and fulfillment.

Perhaps most troubling, even when you take great pains as an adult to escape her reach, you're very likely to carry with you huge reserves of the anger and resentment her control created in you. You may also have a powerful need to exert control in your own life, often by controlling other people. Or, conversely, you may live with the sense that you must always put others' needs ahead of your own. Those are the pervasive marks of having grown up with a controlling mother.

Karen: Trapped and Bullied

When Karen, a twenty-seven-year-old department store sales as- sociate with dark brown hair, came in for her first session, she told me immediately that she was in crisis. Her longtime boyfriend had recently proposed, but at the news of the engagement, her mother, Charlene, had gone on the warpath. Not only had she heaped in- vective on Karen's fiancé, whom she'd never liked, but she'd also threatened to "disown" Karen if she went through with the mar- riage. Karen told me that she was scared of what would happen, but she had enough insight to know she needed to work on getting some distance from her mother's overbearing interference.

Her story poured out when I asked for details about what was going on.

KAREN: "I guess I knew this day would come. By some miracle, I've been with a really great guy for the last two years, but my mother never saw it that way. For one thing, Daniel's Latino and Catholic, and she thinks there's something criminal about both of those. Then there's the fact that he teaches math at an elementary school, and coaches soccer. He's fabulous with kids, and he's got a couple of advanced degrees so he'll have more options with the subjects he can teach. Sounds like a dream, right? But Mom talks down to him, and she refers to him as 'your little gym teacher friend,' which she says with complete disdain. From the beginning she's ridiculed him. No one I like ever measures up in her eyes, especially 'an immigrant,' which is what she calls Daniel.

"The whole time Daniel and I have been together, I've tried to keep them apart and hoped that would be enough. I thought I could just smooth things over. I've always apologized to Daniel about Mom after she's been awful to him, and I tried to just change the subject when she started in with 'How can you be serious about someone like *that*.' It's not worth arguing. But after Daniel proposed, he insisted that we go over to Mom's and show her the ring. I wanted to go alone. . . . I know that's crazy. Anyway, we went. What a disaster.

"Mom didn't even try to be civil. She criticized the ring, and Daniel, and his family. He tried to be polite, but he was steaming. Finally he said, 'I'm sorry you feel that way. Come on, honey, let's get out of here.' And Mom looked at me and said, 'If you go through with this, you're not my daughter anymore. Don't think I don't mean it. You're trying to ruin your life, just like you always have. If you want to defy me, go right ahead. But forget about any help from me with the wedding, or anything else.'

"I froze. I said, 'I'm sorry, Mom,' and Daniel gave me the most pained look and said, 'That's unbelievable. You have nothing to be sorry for. Why are you apologizing?' I didn't know what to say or do. I just stood there crying till he pulled me out the door."

Karen told me that she'd been in a tailspin in the days since, uncertain about what to do or how to deal with the conflict between her mother and Daniel that she'd tried so hard to push away. I asked her if she really thought her mother would cut off contact with her if she didn't get her way.

KAREN: "Yeah, I do. She's been calling and badgering me about breaking up with him. She was at my place until two in the morning the other night, haranguing me about how I'm making a big mistake, and she doesn't want grandkids from 'a person like that.' I told her I didn't want to talk about it, but it didn't get through. She just *assumes* that she gets to call the shots. I told her, 'Mom, please go home,' and she said, 'I'll go home when you tell me you're not going to marry him.'

"I feel so trapped and bullied. I'm sick of her interference, but she's my only family, and awful as she can be sometimes, I don't want to lose her. But I keep backing down from saying anything in Daniel's defense, and I'm sick of that, too. I'm tired of being such a wimp. I feel so disloyal saying this, but my mother has done nothing but control me my entire life. She split up with my dad when I was little and I don't even know where he is now. But she's always had me to boss around. Everything she has ever asked me to do has been to benefit her, not me. She controlled what I wore, what I ate, who I was friends with, even the activities I did when I was younger. And now she's sure she can tell me who to marry."

Dominated Daughters Easily Become Doormats

The foundation for the showdown over Karen's engagement had long been in place. In the mother-daughter twosome left after her divorce, Charlene had assumed the role of authoritarian and boss, grinding Karen down with criticism. She had provided for her daughter's physical needs—and many people would look at that as a sign of love—but she rarely offered affection. Instead, Karen told me, Charlene was often derisive, especially when her friends were around. Young Karen often found herself the object of unwelcome attention when the adults focused on her.

KAREN: "My mom thinks she has a great sense of humor, but really, she was just mean, especially to me. If I didn't want to wear whatever she chose for me, she belittled and made fun of me. I picked out a dress at the store when I was maybe seven or eight, and when I came out of the dressing room she turned to her friend, who was shopping with us, and said, "Whoever thought a daughter of mine would have such trashy taste?" They both laughed at me so hard. I didn't even know why, but the shame was seared into me. I just stood there shaking till she said, 'Go take that thing off!' It was a yellow dress with flowers, and I never wore yellow or flowers after that, even though I love both of them."

Cruel digs and jokes made at a child's expense can cut to the core, and as a girl, Karen faced them often. She learned it wasn't safe to trust her own judgment, and like what she liked, so she protected herself by going along with her mother's choices. After all, Charlene would brook no dissent. She didn't hit or slap; she didn't need to. Her words and tone underlined to Karen that her feelings and preferences didn't matter.

As a result, Karen never had a chance to master one of the most vital life skills: knowing and asking for what *she* wanted.

When controllers tear their daughters down, whether with threats, ridicule, or criticism, they rob them not only of their dignity and self-respect but also of their volition. The controllers' constant criticism destroys young daughters' belief that they're okay, and it makes them extremely vulnerable to control because it erodes the spirit and sense of confidence daughters need to stand up for themselves and go on to live independent lives. Criticism is the fountainhead of control, and control freak mothers discover early that if you tear your daughter down enough, you strip away her ability to be assertive and her will to resist. So they rely on insults and criticism to keep you one-down, hardly missing a beat when you become an adult.

The knocks accelerate whenever controlling mothers feel threatened, as Karen and Daniel saw when they announced their engagement. Charlene panicked, seeing that she was losing control over her daughter's life, and that Karen was shifting her loyalty to her fiancé. In Charlene's mind, the only way to regain the upper hand was to threaten to cut Karen off. That sounds drastic, and counter-intuitive, but Charlene had controlled Karen for so long, she had every reason to believe her daughter would buckle, and that she'd never have to follow through on her threats.

Karen did come close to giving in, she told me.

KAREN: "After all the pressure from Mom, I was so tied in knots I felt physically ill. I told Daniel, 'We're fine like we are. We don't have to rush things.' He just shook his head and said, 'I know exactly what this is about and I'm not going to have this drama hanging over us. I hate seeing your mom walk all over you, and there's no way we can have her dictating the terms of our relationship. You need to get some counseling, do something about this.' I don't think I would be here if he hadn't done that."

I hear that often. Partners or friends may well be the catalyst that pushes you toward change because they see so clearly how unable you are to do anything about the situation.

It's very difficult to move forward from a limbo like Karen's when your healthy instincts to disagree, to say no, to become your own best authority on yourself—all the essential elements of individuation—have been stunted. All the criticism she heard from her mother when she was small turned Karen into, as she put it, "the most people-pleasing person I know." And she felt paralyzed as she tried to figure out how to do the impossible and make both Charlene and Daniel happy. Notably, she never thought to put herself in the equation. It wasn't something she'd had much practice doing.

KAREN: "I avoid conflict at all costs. I'll do pretty much anything I'm asked. But here's the weird part: When I don't please someone in authority—my boss, my mother—I tend to get sick, break out in hives, and become very withdrawn or shut down altogether. I feel terrible, terrible guilt."

Karen became the person who would take jobs no one wanted and be ever available to fill the role of doormat while she ignored her own needs. Raised to be dominated, she had a finely honed ability to abdicate decision making to other people, her mother above all.

The Perfectionists: Holding You to Impossible Standards

Some controlling mothers, like Karen's, seem to turn on their negativity almost on a whim, focusing on their daughters' most recent desires and squashing them, or dispensing cruel put-downs or criticism just because they need to make themselves feel better in the moment. But another variety of controller is far more systematic. These are the perfectionists who seem driven to hold you to a standard that's impossible to meet. They build their households around rules, routines, and drills that are not to be questioned, and they regard anything less than perfection as failure.

Michelle: How Criticism Creates a Critic

Michelle, a thirty-four-year-old graphic artist, told me at our first session that her relationship with her boyfriend, Luke, was on the verge of breaking up. Things had been tense between them for a while, she said, and their last fight had been a big one—he'd taken a few of his things and gone to stay with a friend.

MICHELLE (in tears): "I really thought he was the one, that we'd get married, but he's fed up and he says this is it. I just don't understand why my relationships are so messed up."

I suggested that the two of them come in for a session, and Michelle persuaded Luke, a lanky thirty-year-old video game designer with shaggy brown hair, to join her for an appointment. The two of them came to my office the following week, and I asked Luke to fill me in on what was going on, from his point of view.

LUKE: "Well . . . it seems like the longer we live together—it's been almost a year now—the worse it gets. I had to get away for a while, and right now I'm sleeping on my buddy's couch, but at least I've got a little peace. Michelle's so damn critical. She picks at me for the smallest things. If I've got my stuff scattered around my computer, in my own office in my own home, she goes ballistic. I didn't realize when we first got together how compulsive she was about where she lived and stupid things like the T-shirts I wear. But boy is she ever."

MICHELLE: "Well in my own defense, he's no angel. Yes I have
my faults, but would it kill him to put his socks in the hamper
or put on something decent? How hard would that be? He's
always leaving a dish in the sink, and it's so easy to put it in the
dishwasher. Little things *matter*."

LUKE: "Come on, Michelle. Of all the things in the world to get
obsessed with, why is that such a big, freakin' deal? God, you
sound just like your mother."

Their irritation with each other was clear, and it was apparent
to me that there was more involved than dirty dishes in the sink.
I told Michelle that there seemed to be a whole pattern of picky,
critical, perfectionist behavior on her part that was pushing Luke
away.

MICHELLE: "Oh God. . . . To hear you say that . . . That's my
mom. Picky. So critical. I've always sworn I would never, ever
be like her. And here I am."

Our mothers' imprinting and programming is so pervasive that
it's easy to find ourselves behaving like them without even realizing
it. But patterns can be broken—with effort—once we're aware of
them, I told Luke and Michelle. Were they ready to put in the work
it would take to save their relationship? I asked. They exchanged
glances.

SUSAN: "Think of a container of milk that gets left out on the
counter. Sometimes you can put it back in the refrigerator and
it will still be sweet. But sometimes, it's so far gone it can never
be sweet again. What point do you think your relationship is at
now?"

LUKE (looking at Michelle): "I don't know. I'd like to make it work,
but we don't seem to be able to do it by ourselves. We just have

the same arguments over and over." (He gave her a little smile.)
"But a lot of it *has* been very sweet."

MICHELLE (tearing up): "I don't want to lose him."

I could feel the still-strong connection between them, and I suggested that Michelle and I work together on our own for a while
to get at the roots of that criticism. I told Luke that I thought he
needed to go back home. Many studies on marriage have shown
that the longer a couple is separated, the greater the chances they
won't get back together. It would be tense at first, I told them, but
if Luke could be less reactive and patient for a while, Michelle and
I would be working to modify and extinguish the deeply engrained
patterns of criticism that had been coming between them.

THE MAKING OF A BULLY

Daughters of unloving mothers almost universally promise themselves one thing: If I do nothing else in my life, I will never, *ever*,
turn into my mother. Yet as we've seen, as adults, they often shock
themselves by acting very much the way their mothers did toward
them. Getting to the roots of that behavior was what Michelle and
I focused on in our work together.

Things hadn't been easy at home since Luke had returned, she
told me.

MICHELLE: "He's been calling me on my perfectionism a lot.
    Sometimes I just get defensive or cry or even yell at him. But I
    really notice what I'm saying to him now. For some reason when
    we were here together, I actually got it that I've been acting like
    my mom. It really scares me. . . . I got away from her as soon as
    I could, and we don't see her much now—of course, she really
    hates Luke. *He's* not perfect enough either. But obviously, I took
    her with me. I'm turning into her anyway. . . ."

As she told me about her childhood, we both began to see how many parallels there were between what she'd experienced and what she was living out now with Luke.

MICHELLE: "I was born to parents who never should have had kids. My father was an emotional hostage to his ultrareligious mother and workaholic father. My mother was raised in a horribly dysfunctional home with alcoholic parents and a verbally and physically abusive father. When I was growing up, Mom was a tyrant. There was nothing soft or nurturing at all. She was the strictest mother in the world. She was all about perfection. Perfect, clean home, perfect husband, perfect job, perfect kids. When I was little I would sometimes tell her, 'I'm not perfect' and she'd snap, 'Well, try to be!' That was her sole agenda in life. Her clean home and her job as a paralegal were more important than anything else. And if my sister and I weren't there doing most of the housework, she resented us. My father was constantly working, trying to get a failing restaurant off the ground. Mom resented him, too.

"She was relentless. I worked as hard as I could to at least get good grades, but if I brought home all As and one B-plus, I lost privileges for the B. Mom drilled me in math, but it was more like a military exercise than a math lesson. She sometimes took away my allowance for wrong answers. I had chores like you wouldn't believe, washing and cleaning and lining up the magazines just so on the spotless coffee table. But as far as she was concerned, I could never do enough."

SUSAN: "Maybe that gives you a little clue to what Luke's feeling. If you can remember how awful it felt for you, I think you can get a sense of how it is for him."

MICHELLE: "You mean I've been making him feel like *that*? How could that even happen? My mom was an absolute tyrant."

One of the most common, and distressing, offshoots of a mother's tyrannical control is bullying, which often seems to come at daughters from every direction. Michelle told me that her mother had carefully controlled what she wore to elementary school—"I was the only girl in school who wasn't even allowed to wear pants, let alone jeans," she said—and made her an object of ridicule.

MICHELLE: "It was so awful. The kids made fun of me, and I was so depressed and alone. But the worst thing was that I was taunted and teased and chased by the bullies at school. It was horrible, and Mom never stood up for me. She made it happen with her stupid rules. And she never did a thing to help me—she said I had to learn to be tough. It was the worst time in my life."

It's not hard to see the connection between being bullied at home as a young girl and becoming vulnerable to bullying in the outside world. Pushed to be quiet, uncomplaining, and compliant by a controlling mother, it's natural for a child to take that role at school. She learns to be a target, and she has no skills for protecting herself. She is groomed to be passive—and bullies can tell. Many of my clients have felt the pain of bullying that made them dread going to school.

When children have been through this kind of perfectionistic control, it's not uncommon for them to decide that when they're finally on their own, they'll never again let bullies dominate them. Instead of being pushed around, *they'll* do the pushing. And as adults, they start giving orders about the socks on the floor and the dishes in the sink.

Little of this happens on a conscious level, so gaining awareness of what you're doing is vital if you want to change. It takes motivation and commitment, and the temptation to revert to old ways of being will always be with you, but once you've seen your patterns, you can put in place internal mechanisms that will help you become aware of the impulse to behave like your mother—and no longer have to act on it.

The Sadistic Controllers

Taken to extremes, control can become out-and-out cruelty, with a mother's rules and standards shifting constantly and harsh punishments meted out for no reason that a daughter can anticipate or understand. Cruel controllers are more than bullies. At the far end of the spectrum, some of them exhibit strong elements of sadism. They seem to derive some kind of warped pleasure from humiliating and thwarting their daughters or seeing them suffer.

Living with a sadistic mother, daughters are constantly off balance, shamed, and often afraid, and long after they leave home, they often keep their fight-or-flight responses close to the surface. The urge to run, or to come out fighting, is a survival strategy that has served them so well, sometimes they hardly realize there are other ways to live.

Samantha: A Legacy of Anger, Turned Inward and Out

Samantha, an elegant twenty-nine-year-old African American who manages a sales team for a large pharmaceutical company, started our first session by telling me she'd had a confrontation at work that deeply disturbed her.

SAMANTHA: "We have a new area manager—we're supposed to be peers—and she's really working my nerves. I mean, she's good, but she acts like she's queen and the rest of us are nothing. At a team meeting, she came really close to trashing me, setting *me* up as the problem when it's actually she who's been so

disruptive and bad for morale. I pride myself on being cool
and calm and never letting people see how I really feel. I'm
not cold, just professional. But after that meeting, something
snapped. I lost it. I was okay while she was talking about me.
My face got hot, but I didn't say anything. It was at least the
second or third time she'd done it. I wanted to just stay cool
because she's new, and people like her. But in the parking lot at
the end of the day, she made some joke about me and I just lit
into her. It was pure rage, and to tell you the truth, I was out of
control. I yelled like a crazy person. . . . And I got this almost
high feeling. I know it scared her, but it was pretty scary for
me, too."

I told Samantha that blowing up that way might feel good for
a moment, but as she well knew, the consequences make your life
worse, not better. A lot of people think that if they yell, they're
standing up for themselves; however, not only doesn't it solve anything, but it also makes you lose your dignity and your credibility.
There are so many better ways to deal with anger.

SAMANTHA: "I know. Yelling in general freaks me out. I grew up
     with it, I hate it, and I just shut down when people raise their
     voices. I hold things in for a long time . . . and then I explode."

It's common for people to try to stay safe by shrinking away from
someone who's yelling. For children, especially, it makes sense to
shut down and try to disappear—to become less of a target. But
the strong emotions they feel don't go away. Samantha had clear
memories of how terrified she'd been as a child when her mother
yelled at her.

SAMANTHA: "My mother could be . . . a real bitch. I'm sorry but
     don't think there's another word for it. She was full of rage. I

don't know why, really. There was plenty of money—my dad was in-house counsel for a biotech company, and she works as a lawyer for a utilities company. They're both brilliant, and I think they expected me to be on par with them from the time I was little.

"I remember when I was maybe three years old and Mom was trying to teach me the *ABC*'s. Most mothers have you sing the alphabet song and make it a game, but my mother didn't believe in that. She came into my room and told me to say it, say it. Again! Again! I didn't know my letters by heart, and she screamed at me so loudly I was terrified. I can still hear her voice in my head."

As Samantha got older, her mother's irrational control and cruelty took on new dimensions.

SAMANTHA: "I was always tall for my age. When I was fourteen, I made it onto the basketball team. That was my dream. We were really good, and we made it to a tournament in Boston. I was scheduled to go with my friends. I was so excited—it was going to be such a fun time. I saved every dollar I got from babysitting and bought my plane ticket. But at the last minute, my mother told me I couldn't go because my grades weren't good enough. I'd gotten a C on a quiz for the first time. It wasn't even going to count! But according to her, I was going to flunk out. She said I needed the time to study, 'not play.' . . .

"I remember sitting in my room watching the clock, hoping until the last minute she would change her mind. I still remember the moment when I knew I was going to miss the plane. I called my coach to say my mother wouldn't let me go. He was really upset and asked to talk to Mom, but she wouldn't get on the phone. God, Susan—there was absolutely no plausible reason for her not letting me go! My grades were fine—I had a lot of As and Bs! She just wanted to flaunt her power over me. . . . She could take away anything she wanted."

Some mothers get a kind of warped sense of satisfaction from depriving a young daughter of something she wants. And like so many daughters of sadistic controllers, Samantha fantasized about escaping.

SAMANTHA: "When I was in junior high, I practiced packing a little backpack with everything I needed to run away. I timed myself. I could be ready in ten minutes. I don't know where I thought I would actually go, but I needed to believe I could."

Her real escape, though, came when she was older, and it didn't even require leaving home.

THE REBEL ROUTE

It's extremely common for daughters like Samantha to "take control" of their lives and attempt to squirm out of their mothers' rigid constrictions, rules, and punishments by rebelling.

SAMANTHA: "My mother thought she could make me do anything she wanted, but by ninth grade, I figured out that the one thing even *she* couldn't really control was my body. One of my salvations came when I was able to start dating and sleeping with boys. I had to sneak out to do it, but it was worth the risk. I figured out that I could use my body to have some sense of control over myself. I started bingeing and purging then, too. I was a serious bulimic for a long time under her roof, and she never really noticed, even when I stopped eating and looked almost anorexic.

"I took honors classes and graduated early so I could

get out of the house, but after all that, I think I spent a considerable amount of time in my life in college and after trying to hurt myself. I felt guilty and depressed a lot of the time, and the only things that really helped were having sex, getting drunk, or getting stoned. Purging was good, too. I mostly wanted to check out. I hated myself, I hated my life. But I had a friend who was going to AA, and she asked me to come to a meeting with her one night. And everything changed for me. If it hadn't been for her, I don't know what would've happened."

Sadly, many daughters of unloving mothers get their first taste of freedom and often destroy it by acting out self-destructively. Whether it's with alcohol, drugs, food, sex, or all of the above, the rebel often degrades herself in a fruitless attempt to prove that her mother no longer controls her. When we are thwarted, frustrated, and punished way out of proportion to what we've done, it's inevitable that enormous anger builds inside us.

This anger may feel difficult and uncomfortable, but it can become a very good catalyst for change. When it's not expressed appropriately, though, it can be extremely destructive. Often the anger turns into depression, which can build to the point where a daughter will do almost anything to escape the chaotic feelings inside. Some of my clients told me that they have even considered suicide. It's a cycle of anger and despair that may well persist through adulthood.

Self-destructive rebellion isn't freedom, because the rebels' choices aren't based on building up their own confidence and self-respect. They *can't* be truly free. Instead they still have their mothers in their heads and act out in ways aimed to shock and upset them. They've never really learned how to construct a life that reflects their own desires. Ironically, their mothers are still controlling them.

What's Driving the Controlling Mother?

As I think about the many controlling mothers my clients have had, certain facts emerge clearly. These mothers seem to be very displeased with their lives. They may have come from homes in which they were, themselves, controlled and belittled by their parents. They may be controlled and put down by their husbands or bosses. Their roles or freedom may have been limited in ways they internally chafe against, but feel helpless to change. Anger, bitterness, frustration, and disappointment may well be swirling beneath a tight smile. Without some sense of empowerment, they feel lost.

Whatever the roots of their need to control, these mothers will flex their control muscle by belittling and criticizing your appearance, choice of schools, job, partners, wedding preparations. Like so many other mothers who can't love, the controllers make the most of your every vulnerability.

But the control that often has the most far-reaching impact on your life comes from the patterns, reactions, and expectations your mother has implanted so successfully in you—even if you think you've pushed her away.

If you're struggling with people-pleasing, perfectionism, a tendency to bully or be bullied, or any of the other painful behaviors we've seen in this chapter, let me assure you that these are learned behaviors—and you can *un*learn them.

Chapter 5

Mothers Who Need Mothering

ᴪ

"I depend on you to take care of everything."

A mother can't teach her daughter to navigate life if she takes to bed with a bag of M&M's every afternoon and shuts the door, or if she's passed out on the couch when it's time to get the kids up and ready for school. She may not be available to cook dinner, take care of younger children, or look after herself. Whether she is depressed, alcoholic or addicted, or infantile, when she needs much more mothering than she can give, her daughter finds herself taking on the role of parent, protector, and confidante.

For a young girl, few things are more distressing than detecting that "something's wrong with Mom." And with these mothers, there's definitely something wrong.

Mothers who need mothering frequently withdraw into their own world, abandoning their role as caretaker. They may be at home, but they're rarely present enough to notice your accomplishments or wipe away your tears after a disappointment. Instead, they spend their days sleeping, complaining, watching TV, drinking; and daughters who've known nothing else rarely recognize a poignant truth—that they are essentially unmothered.

The mothers you'll see in this chapter are MIA—missing in action—they've simply checked out, putting whatever energy they have into their own survival, with very little left for tending to their daughters' well-being.

Most of their daughters grow up feeling tremendous pity for them and believing that it's their job to "make everything all better"—whatever that takes. Girls who are forced into this kind of role reversal often take pride in being called "so grown-up," "responsible," and "wise beyond your years." But they've essentially been robbed of the chance to have a healthy childhood.

As adults, many of them pride themselves on being cool, capable, and able to take charge. They have a lifetime of practice shouldering burdens and taking on responsibilities that belong to others. It's second nature for them to become the person others turn to for support and encouragement, and they know exactly how to be a vehicle for other people's survival, success, and happiness. But when it comes to their own needs, they come up empty. They've rarely learned to put themselves, their dreams, and their own joy at the center of their lives. What they've mastered instead is the all-consuming art of being a caretaker.

Telltale Signs That You Grew Up as a "Little Adult"

It's often so difficult for adult daughters to step back and see how they were put into the adult helper role. To help you recognize if this dynamic echoes your experience, I've created a pair of checklists to help you identify how mothering your mother shaped and influenced a significant part of your life.

When you were a child did you:

- Believe that your most important job in life was to solve your mother's problems or ease her pain—no matter what the cost to you?

- Ignore your own feelings and pay attention only to what she wanted and how she felt?
- Protect her from the consequences of her behavior?
- Lie or cover up for her?
- Defend her when anyone said anything bad about her?
- Think that your good feelings about yourself depended on her approval?
- Have to keep her behavior secret from your friends?

As an adult, do these statements ring true for you:

- I will do anything to avoid upsetting my mother, and the other adults in my life.
- I can't stand it if I feel I've let anyone down.
- I am a perfectionist, and I blame myself for everything that goes wrong.
- I'm the only person I can really count on. I have to do things myself.
- People like me not for myself but for what I can do for them.
- I have to be strong all the time. If I need anything or ask for help, it means I'm weak.
- I should be able to solve every problem.
- When everyone else is taken care of, I can finally have what I want.
- I feel angry, unappreciated, and used much of the time, but I push these feelings deep inside myself.

The cost of growing up as a "little adult" who never had the freedom to be a child is high. If your entire value as a child came from being a caretaker, you never were able to develop your individual self, enjoy the freedom of imaginative play, or learn to let down your guard and be spontaneous. There was little time or support for asking "What can I be?" or trying on different identities on your way to finding a satisfying path of your own. Instead, you trained your focus on your mother, becoming an expert in her needs rather than your own, and vigilantly trying to anticipate difficulties and step in to resolve them.

But there is a cruel twist built into the role-reversal dynamic: It's always a setup for failure. A young child doesn't have the power to solve her mother's problems—only her mother can do that. Even the biggest smile or sacrifice a child can offer can't change Mom. But the daughter feels compelled to try. And when her efforts fall short, she can't help feeling inadequate and ashamed. Young daughters deal with those feelings by resolving that when they're grown, they'll "get it right," and as adults, they work tirelessly to do just that. They do too much for other people, give too much, help too much. It's what psychologists call a repetition compulsion: the need to repeat old behavioral patterns with the hope of getting different results in the present than you got in the past.

When that compulsion drives you, your life can look like an endless series of burdens to be lifted from others, a treadmill of problems to be solved. Joy, lightheartedness, and fun go missing. And it becomes difficult to distinguish love from pity, or to believe that love relationships can be reciprocal—free of the need to rescue.

Allison: Falling for "Fixer-Uppers"

Allison, a willowy forty-four-year-old yoga instructor with her own studio, told me she had long suffered from depression, and she wondered if she'd ever be able to have a satisfying, loving relationship with a partner. She had a history of getting involved with men she had to take care of, and she made her first appointment with me fresh from a fight with her most recent partner, Tom, the man she'd been living with for eight months. I asked her to tell me about what happened.

ALLISON: "You know how they say opposites attract? I guess that's what happened to me. I've always been so careful, figuring out the grand plan, doing things by the book, being the good girl. So when I met Tom, it was like, 'Wow. Life can be way more fun than this.' He was working part-time as a waiter, doing photography the rest of the time, and the tiny place he was living in was covered with these photos he'd blown up and painted. He didn't have any money, and he didn't care. He was so creative. I'd never met anyone like him—kind of a bad boy on a motorcycle. I really fell for him, and I was in awe of his talent and all that freedom he had. His friends were arty and wild—it was a whole different world for me.

"We started living together, at my place, since there was no room for anything in his, and it was great at first. He filled up the house with his photos, and sometimes there would be a party going on when I got home from the studio. I admired the way he just worked a few days a week so he could stay true to his art, and I knew he could do so much better if he had fancier equipment. So I got it for him. I've never seen him so happy, and I got really excited for him. I really thought that his work could take off and he could be a top-flight photographer."

SUSAN: "Okay, let's see. You move him into your house, he's throwing parties there, he only works a few days a week, and you're buying him all this fancy equipment. I can imagine how *that* worked out."

ALLISON: "Yeah, not so great. . . . All the stuff I bought was like new toys for him, and after a few months he lost interest. He didn't even pretend to look for photography work. I came home early the other day and he was smoking pot with all the windows open and watching TV, like he does when he knows

I'm going to be out. His camera was right next to a full ashtray, sitting on a sticky table. He just can't get motivated to take the next step. That's what the fight was about. He yelled at me and said, 'Okay, fine! I'll just go wait tables again.' I feel so ripped off and disappointed! Tom is so damn dependent.

"I feel like I married my mother. We're not married, but you know what I mean. He's just like her. I always get involved with men I want to nurture and save. Not the together guys, but the ones who have, you know, 'potential'"—she made quote marks around the word with her fingers—"and just 'need to be loved.'

"That's just the way I am. I've always been the together one. My mother said I was an old soul, but it wasn't true. I guess I had to grow up fast because of my family. My mother leaned on me a lot when I was a kid."

I asked Allison to tell me more about the way she'd grown up, and as she did, it became clear how early she had learned the care-taking behavior that was shaping her life with Tom.

ALLISON: "My mother was a stay-at-home mom. My dad had a terrible temper, and they fought whenever they were together. It saved us that he was always away on business. I was raised to lie to my dad—there was always lying and deceit to keep his temper at bay, and walking on eggshells to keep him from exploding. My mom thought he was sleeping around, and he probably was. She hated him for that. But she was so helpless. She wanted to leave him, but she was afraid she couldn't make it on her own with me and my little brother and sister. So she stayed. And I heard all about it. I realize now I was exposed to too much information at a very early age."

SUSAN: "It sure sounds that way. You were supposed to be out with your friends having fun. What were you supposed to do with all that information?"

ALLISON: "I don't know. And even today, she asks me if she should get a divorce. Then she told me she was only staying with him because of us, so it was our fault. I don't know how many times I've said, 'Just leave him!' But I've given up. Nothing's ever going to change. Her kids are grown up now, but she's still paralyzed. She just can't bring herself to do anything but complain. I'm so frustrated I want to scream, but it hurts so much to see her suffer. I still feel like I have to cheer her up and somehow patch things together. When I was a kid, if anything was going to happen to keep us going and feeling even a little bit like a family, I had to do it myself. Cook. Clean. Buy the Christmas tree and remember the kids' birthday presents. I did everything. Just like with Tom. God, Susan, I'm so tired of doing everything. . . . When is somebody going to take care of *me*?"

At that point, Allison broke into tears and cried for a few moments, and as she wiped her eyes, she softly said, "I'm sorry." Like so many women, Allison found it necessary to apologize for crying, as if she'd done something wrong.

I told Allison that she had every right to cry and be upset. The guy she fell for turned out to be needy and irresponsible. And there was a lot of grieving to do for the girl she'd been, who was asked to take on the role of a mini adult and take care of not only her mother, but sometimes a whole household. That's a pretty staggering load to place on the very narrow shoulders of an eight- or ten-year-old who lives with the knowledge that if she wants to complain or exult or just be eight, there's no one at home to turn to.

Allison didn't linger in her sadness, though. She quickly composed herself and, as she had done all her life, did her best to absolve her mother of responsibility.

ALLISON: "To be fair, Susan, it wasn't her fault. She really did
have a terrible marriage and a terrible life. She was so sad so
much of the time. I hated to see her like that."

It was only natural for Allison to turn once more to her deep
well of sympathy and lavish it on her mother one more time.

Depression Doesn't Erase Her Responsibility to You

Though I didn't see Allison's mother, Joanna, I think it's reasonable to believe that she was beset by the demon of depression. In
fact, I don't think it's going too far to say that most mothers who
need mothering are beset by the same demon. Depression exhausts
and paralyzes them, decimating their ability to nurture, or guide,
or comfort. There may be moments—even short, good periods in
their lives—when they appear to be available and caring, but their
need to be taken care of overwhelms everything else.

These mothers are caught in a dark spiral, their sense of possibility dimmed by their illness. Allison, like so many daughters,
grew up with her mother's hopelessness, steady as a heartbeat, creating a heavy atmosphere of pity and sadness with words like:

• Life is terrible.
• I wish I'd never been born.
• What have I done with my life?
• Why did I marry your father?
• I don't know what to do. I've screwed up my life.

Depression robs these mothers of themselves and distorts their
decision making. Their condition is the result of some combination
of genetic factors, physiological factors, and unhappy life circumstances. A depressed mother is ill, and she's suffering.

However, she *is* an adult, and she's responsible for taking steps to change her situation and improve her life. That's true for all adults. It's not a suggestion, it's a *mandate* for a mother to help herself so she can adequately care for her children. Even if, like Joanna, she is consumed by her own fears.

The resources for treating depression have improved enormously over the past several decades. Antidepressants have been effective for vast numbers of people, and there are many alternatives for addressing this debilitating condition. But so many mothers like Joanna, who had all the marks of being severely depressed, often back away from getting help and surrender to the victim role.

At the end of our first session, Allison told me that her mother resisted every suggestion that she seek treatment.

ALLISON: "I've tried, Susan. I just tell her, 'You know, Mom, there are people out there who can help you. Your doctor, a counselor.' But she won't even consider it. She fights me! 'How can you say that? I'm not the one with the problem. I haven't done anything wrong—it's your father. Why should I get help? I'm not crazy. Your father needs to stop yelling, that's all. I'm not the one who needs counseling.'"

In cogent moments, a depressed mother may find enough energy to notice her daughter and offer a feeble "You're so cute" or "You're so sweet." But that doesn't make up for the lack of basic, core validation and bonding that all daughters need so much. Instead what the daughter hears most often is: "You're so wonderful *for helping me.*" Not for *being* who she is, with all her uniqueness and value.

I have great compassion for mothers living with the heavy darkness of depression. But they are still accountable for the care of their young daughters. And I believe they also need to acknowledge the responsibility they bear for the pain they cause when abdicating their role causes patterns of caretaking to take hold in their daughters' lives.

There was a clear line from the way Allison rescued her mother and Allison's impulse to "adopt" Tom. The pleasure she got from seeing him light up at her gifts led her to believe that now she was at least partially making up for her childhood inability to save her mother. It was the repetition compulsion at work, but as I told Allison, together we would break the cycle so she could finally focus on what *she* needed and wanted.

A Legacy of Depression Isn't Doom

I want to assure you that if you struggle with depression, as so many daughters of depressed mothers do, you're not doomed to handle it the way your mother did. That was one of Allison's great concerns. "I have to admit that sometimes I feel like I caught the depression bug," she told me. "I look at my life and my relationships and sometimes I feel like giving up. I've really had trouble with depression. I don't want to become like my mother."

Daughters who've grown up with depressed mothers not only often have a propensity toward the condition in their genes and brain chemistry, but also often battle the blows to self-worth and self-esteem that come with growing up unmothered.

But as I told Allison, there's a huge difference between you and your mother. You're not slipping into the victim role and saying, "Poor me." You are trying to change.

Jody: Living with a Mother's Drinking, Drug Abuse, and Depression

Role reversal and its damaging effects are pronounced when mothers are addicted to alcohol or drugs. Chaos and crisis are part of an addicted mother's everyday life, and for a daughter, that means even the quietest day has the potential for explosive drama. Jody contacted me by e-mail after a family gathering spun out of control.

FROM JODY'S NOTE: "Dr. Forward, I need to see you. I need to separate from my alcoholic mother, who's always been controlling and critical. I've had it with her. . . . I have lived to please her for thirty-two years and I can't do it anymore. . . . Having her in my life is hurting my marriage and it's making me miserable. HELP!"

At our first session, Jody filled me in on what prompted her to contact me. An athletic-looking blonde, she grew up an only child, was thirty-eight and married, and taught special needs children at an elementary school.

JODY: "It was just a week ago, Thanksgiving in fact, that finally did it. Mom ruined everything. I have lots to be thankful for, a great husband and a beautiful new baby, and things should be wonderful. But they never are with Mom around, and this was the final straw. We were watching the Macy's parade and then football, eating and playing with the baby. But out of the corner of my eye I was watching Mom reach for the wine, counting the glasses she had. I would walk by and move the bottle—it's an old habit—but my brother-in-law would refill her glass. I could've killed him.

"Mom was starting to get loud and a little slurry. She sat down next to my aunt and knocked over her glass of red wine. They mopped up everything, and my aunt grabbed Mom's glass and said, 'Margaret, I think you've had enough.'

"Mom was furious. 'You want to know why I drink so much?' she yelled. 'I'll tell you why I drink. *This* is why I

drink.' She pointed at *me*. As if it was my fault! Then she said,
'Selfish, know-it-all girl with that counseling degree. That's a
joke. She's sick, sick in the head.'

"It was just unbelievable. I wanted the floor to open up and
swallow me. That was it for me. Something has to change. I
hate her drinking, her pill popping. She takes pills to go to
sleep, pills to wake up. And she's so selfish and depressed. . . ."

Jody was overflowing with anger, and I knew she needed to vent
and have me hear her.

JODY: "You'd think I'd be used to it by now. I don't remember
when she was anything but a drunk. So many times when I
needed her she was too drunk to be there for me. She moved
me from home to home and dated so many men, then brought
them into my life and acted like they would be a dad to me
and we would be a real family. She left me alone a lot to work,
date, or just do her own thing. And if she was home, she was
drinking or drunk or passed out."

Jody's mother, Margaret, was almost never available for her
daughter, and over several sessions, Jody gave many examples of
how she'd been neglected, and what a remarkable amount of responsibility and stress she'd assumed so young.

JODY: "I remember how I'd wait for her to come home from work
when I was in fourth grade or so. I'd have dinner ready for
her, and she'd tell me about work. She didn't get along with
her boss. I was so afraid she'd lose her job, and then what
would we do? But she didn't seem to worry about anything.
As soon as we were done eating, she'd go to her room with the
newspaper and prop herself up on the bed with the TV on. She

had a bottle of Scotch on the nightstand, and her glass, and she'd 'read' and sip her drink. Usually she wouldn't even take her shoes off. She'd fall asleep with the TV blaring, and a lit cigarette in her hand. I'd go in and take the cigarette before it burned anything. Then I'd cover her up and dump out the rest of the bottle, for all the good that ever did.

"I'd go wash the dishes and turn on the TV in the living room so I could have some company doing my homework. It was so lonely, Susan. If it wasn't for my friends at school, I would've been the loneliest kid in the world. I guess I was just supposed to raise myself. But I prayed a lot, and it helped me feel closer to God."

Daughters of addicted mothers can't tell friends and teachers what's going on at home. When they go to friends' houses and see that everyone doesn't live the way they do, they realize there's one more secret to keep, and they live with a sense of shame, a sense of being alien or different, even as they become masters of putting on a good front.

Lured by Drugs, Alcohol—and Helping

All the way through junior high, Jody took care not to call attention to herself except with decent grades and neat outfits that she carefully washed and pressed. But in high school, as her friends were negotiating curfews and dating restrictions, she realized that there was an "advantage" to being unsupervised. She could do anything she wanted, and her mother was generally too out of it to say a thing. There was little in the way of meaningful rules, discipline, or boundaries. Her mother didn't raise an eyebrow when young men in their twenties showed up at the door to pick up teenaged Jody. Often Margaret wasn't even home.

Jody began experimenting with drugs and alcohol on her own when she was in her early teens, pouring drinks and occasionally taking pills from her mother's purse. There is a huge danger of children of alcoholics becoming alcoholics themselves—studies say the odds can be as high as 50 percent. But fortunately Jody was able to pull herself back from danger.

JODY: "Yeah, I turned to alcohol to cope. I didn't become an alcoholic by the grace of God and my own realization that I was in trouble. I haven't had more than an occasional glass of wine for more than fifteen years."

Jody credits a teacher in her junior year of high school with helping her begin to turn her life around.

JODY: "I loved my psychology class and the way we got to talk about real stuff in people's lives. It was the only class I did well in, and sometimes I'd go by at lunch and talk to the teacher. She thought I was smart, and she told me I should study psychology in college. It was the first time I could even imagine going to college. She said she and the school counselor would even help me apply for scholarships and financial aid. You don't know what it meant to have someone who believed in me. She got me into volunteering with special needs kids, and it opened a whole new world for me. I loved the kids, and I could just calm down and have something to do besides party. It made me feel so good about myself."

I wasn't surprised that Jody had been so drawn to that field. Adult daughters of alcoholics and addicts are quite likely to become caretakers by profession. They often gravitate to careers in medicine, especially nursing, as well as social work and counseling—it's a very adaptive way to use their drive to take care of other people.

I could hear the sense of failure that fuels the repetition compulsion when Jody told me: "I had no one else but my mom growing up, *so I had to fix her to make her happier. And she never really was.* I felt sad in my heart, like it was always hurting." I'm certain that Jody's gift for teaching, and the satisfaction she gets from her students, helped assuage that sadness, and I'm sure it was helping quiet the unconscious sense that she'd let her mother down. She'd built a workable life for herself.

But her mother was still acting up, still drinking. And now, helped by her outrage at Margaret's recent behavior and her concern for her new baby, she was finally coming to terms with the fact that it wasn't her responsibility at all to fix her mother.

Finding the Courage to Fix Your Life, Not Hers

In all those years of caretaking and neglect, Jody had felt a lot of anger, she told me, "but I was never allowed to be angry for long. We always had to make up quickly, no matter what she pulled. Because our family was just the two of us. And I had to do whatever it took to keep us from sinking."

But Jody had a family of her own to think of now, and as Margaret became progressively irresponsible, Jody was beginning to let her long-simmering anger out, and to see her mother in a more objective light.

JODY: "There's a drawer full of unpaid bills in the kitchen, and
    I think I saw a warning that the electricity was going to be
    turned off. When she calls, she makes her voice sound all
    depressed, expecting me to run right over. One of the first
    things she said when I told her I was getting my master's in
    counseling was, 'Oh good! Now you can fix me!'

"Well, guess what? I've finally had enough. I've tried everything I can think of to live my life and still remain in contact with her and it doesn't work. She can't stop saying hurtful things. She won't stop drinking. What I want now more than anything is to live my own life, for her to leave me alone. She can do whatever she wants—stay in her room, drink, get depressed. I don't care! I just want her out of my life. . . . But . . . how can I just abandon her? She'll die, and then how can I handle the guilt?"

Jody stared into her lap, looking almost physically deflated.

SUSAN: "You have a big responsibility, Jody—to yourself. You've done everything you can for your mother, and from what I can see, she will not do anything for herself."

I asked Jody if she'd talked to her mother about getting help— AA or working with an addiction specialist.

JODY: "Oh, according to her she's 'not an alcoholic.' She somehow still has her job, she's not out on the street yet, and I guess that means she doesn't have a drinking problem. It's always everybody else's fault. She drinks because of me. Right."

Alcoholics like Margaret typically project the blame for their drinking onto whatever seems convenient, I told Jody—the people who are closest to them, world events, the weather. They need just the slightest excuse.

JODY: "I keep telling myself all that, Susan, and they tell me that in Al-Anon. But even when I'm the most furious with her, I feel like she's my . . . my child. And how do I abandon my child?"

Even with all the clarity that they may get from their adult perspective, many daughters have tremendous ambivalence about breaking off from a mother who has essentially morphed into a helpless and needy child. Their feeling of obligation to her is so ancient and unquestioned that it can pierce their anger and their healthy self-protective urges in an instant. Breaking away from all that requires unlearning layers of old responses and setting priorities that have nothing to do with getting sucked into your mother's disasters and depression.

For Jody, the obvious priority was her baby, a real child who needed her and who was actually helpless and dependent. I knew how committed she was to being a strong and healthy mother, and how much she wanted to be there for her daughter in a way that her mother couldn't be there for her. But it takes a lot of physical and emotional energy to be a good mother, and emotional energy supplies aren't infinite. If you have children, you can't keep dissipating your emotional resources by going back to rescue your mother. You have a responsibility to yourself, your partner if you have one, and to the children. *Your mother has to take responsibility for herself.*

When addiction is in the picture, the one certainty is that the addict's substance of choice will take increasing amounts of her attention and resources, whether the "substance" is alcohol, prescription or illegal drugs, food, gambling, or sex. Pulling away from her is the only way to transform the effects her condition has had on you—and that requires disengaging from the behaviors you've been taught: the secret-keeping, the rescuing, the hypervigilance. You'll have to stop doing the kinds of things you take for granted, as Jody did, counting how many glasses of wine your mother has been drinking, for instance, instead of playing with your baby. It's hard work, but it's the only way to keep from passing all the pain of your childhood on to a new generation—or continuing to carry it inside.

You Lost Your Childhood—and It Still Hurts

A daughter like Jody or Allison does her best to make her life seem
and feel "normal" when she's young, covering up the evidence of
her mother's depression, drinking, drug abuse, or neglect. She cares
for her siblings. She cooks, she cleans. If her mother's husband or
boyfriend turns violent, she's the one who puts the antibacterial
cream on her mother's wounds or calls the police. She carries a
horrendously heavy emotional load.

If you are a woman who grew up with a mother who abdicated
her maternal role, you may have taken a great deal of satisfaction
from being needed. Some of that behavior looks noble on the surface, but you've paid dearly for it. You got cheated out of a childhood. You have a right to be both sad and angry about that.

Chapter 6

Mothers Who Neglect, Betray, and Batter

⚑

"You're always causing trouble."

Just like sea turtles who deposit their eggs in the sand and then go back to the sea, some mothers disappear emotionally almost as soon as they've given birth to their daughters. Unavailable, distant, and cold, they may be physically present, but they look right through their little girls, preoccupied with their own needs.

Self-centeredness is common to all the mothers we've seen, but the mothers at this end of the continuum are so disturbed that they neglect their daughters' basic emotional needs, and sometimes their physical ones. So incapable of caring are they that they put the lie to the assumption that bonding is an intrinsic part of motherhood. Women like this treat their daughters like objects, resenting them, blaming them for life's dissatisfactions, withholding even the smallest kindness, and, in the worst cases, failing to protect them from predators and abusers—or becoming abusers themselves.

These mothers who emotionally abandon, betray, and batter are mothers in name only. And they leave in their wake daughters who are fearful, angry, ravenous for affection, and forever struggling to find their own way.

Emily: The Invisible Daughter

Emily, a thirty-six-year-old comptroller for an architectural firm, contacted me for help with her two-year relationship. At her job, she felt competent and respected, she told me, but her closeness with Josh, who had a small importing business, seemed to be slipping away.

EMILY: "I have good friends, I'm making good money. But I'm so miserable at home. I thought Josh was sexy and exciting when we got together, and I thought he wanted kids. I really want a baby, and I can hear the clock ticking. But everything went bad between us. Josh keeps everything to himself, and he's so withdrawn—we're living together but I feel so alone. He's always on the computer, and even when we go out, he says so little, retreats into his phone. He leaves me starved for love. The sick thing is that it feels so familiar it's almost comfortable."

I asked Emily why this was familiar to her.

EMILY: "This is so hard to say, but my mom was like that—so distant and cold. I . . . didn't feel like she wanted me around."

The loneliness and distance she felt with Josh, she told me, was very much like what she remembered from being a child.

EMILY: "My mother had me, but she never hugged me or told me she loved me. When she did talk to me, it was to tell me what I had done wrong and what a burden I was to her. Once she even said, 'I wish you'd never been born.'"
SUSAN: "Oh, Emily. I'm so sorry that happened to you. 'I wish you'd never been born' is the cruelest and most wounding thing a mother could say to a child."

Emily teared up. "Thank you," she said softly. "That's the first time anyone's ever really heard me."

We sat quietly for a moment, and then I asked Emily if she got any affection from her father.

EMILY: "My father was out of it most of the time. He worked really long hours. Looking back, I think he did everything he could to avoid her. So I never got any guidance, no teaching, no love or support. Why did they even have me if they didn't want me?"

Emily believed she had to be the only person rejected so dramatically by her mother, but I reassured her that sadly, it's a story I've heard all too often. Many daughters have told me they've been ignored, made to feel invisible and unwanted by mothers who starved them of attention, touch, warmth, and support.

Mothers like Emily's look at their young daughters seeing only "mess" or "bother" or a disruption of the fantasies and plans they had for themselves. In their preferred vision of life, they're unencumbered by a child's needs. And to them, the sweet face of their little girl—an innocent being who loves them unconditionally—is scarcely visible.

We look at these mothers and wonder: How can they be so untouched, so unmoved, so callous toward a helpless, hapless child who is completely dependent on them for emotional sustenance that is as essential and life-giving as milk?

What creates these situations? The reasons are many and varied. We have to assume that a mother who is so cold and uncaring must have been severely traumatized herself. She may have been rejected, or grown up in a loveless household and never learned even the rudimentary aspects of tenderness, empathy, or giving. That kind of trauma doesn't go away by itself.

When these women become adults, they often get caught up in the social pressure to have children. Some give in to a husband's desire to have a baby when it's not really what they want themselves. Or they unwittingly become pregnant and feel compelled by their moral or religious beliefs to become mothers, despite their own misgivings. Then, when the baby arrives, they suddenly have to face the reality that having a baby dramatically changes a woman's life, demanding attention she may not know how to give.

A woman like Emily's mother almost certainly was a stranger to love. Without a spark of love to soften her fears and frustrations as she navigates the new world of motherhood, such a woman fills with anger, and her daughter becomes the scapegoat for her discontent, boredom, or sense of helplessness about her life. She wants that child out of her sight.

The Scars of Feeling Unwanted

The kind of emotional abandonment that Emily experienced may seem far less dramatic than, say, a mother leaving a baby on a church doorstep or driving off in the middle of the night for a new life with another man, but it's every bit as confusing, disorienting, and scarring.

EMILY: "I didn't get to feel safe or be a child. I wasn't given any safety net. No teaching, no instructions, no structure, no love or support in any area. I was so ill equipped to handle life. I didn't know how to do basic things. I could never count on my mom for anything. She never made me feel like I was a daughter. I never felt like I was a treasure to her. I was just something she had to deal with when it suited her.

"I felt so abandoned. When I got my first period I didn't know what was happening and I went to my mother. Her response was, 'Handle it yourself.'"

Emily decided early that negative attention was better than no attention.

EMILY: "At least when my mother had to come to school because I was caught cheating on a test or kissing a boy in the hall, I could pretend she cared about me. I wound up getting into a lot of trouble, but if I wasn't in trouble I was invisible."

Invisible. It's a word I've heard so often from daughters like Emily. Her mother essentially erased her, and she had such a hunger for love that she'd do anything to get it. She never learned that she could be loved for herself.

EMILY: "I made bad choices with men. I would give up my money, my success, my plans—anything—to get someone to love me or want to be with me. I longed for people to take care of me, and it never worked out right. They all turn out like Josh—so great at the beginning and then they wind up pulling away, if they were ever really there." (She began to cry softly.)
   "I just don't feel I'm good enough for a good relationship or a good guy. Sometimes I wonder what I'd be like if only I'd had a normal mother who actually gave a damn about me."
SUSAN: "Emily, I want to help you move ahead, and to do that, you can't stay stuck in 'if only,' because 'if onlys' keep you trapped in longing and fantasy and wishful thinking."

I told her that we'd work on two tracks, exploring both her relationship, which was the current crisis, and her childhood. She could learn new ways of being and feeling and becoming visible, whether she stayed with Josh or not.

The Mother Who Fails to Protect

Just as a lioness will battle to the death any creature that threatens her cubs, a loving mother must do no less. Of all the responsibilities that a mother must fulfill if her daughter is to thrive, perhaps the greatest is protection. A mother who *knowingly* fails to protect her daughter from harm or from physical or sexual abuse at the hands of a father, stepfather, or anyone else is guilty of aiding and abetting the perpetrator. Emotional abandonment takes on traumatic and danger-ous facets when she betrays her daughter by standing by and allowing physical harm to befall her.

Fearful, passive, and destructively self-serving, some mothers will permit their daughters to be pummeled or sexually molested rather than confront the abuser and take the risk of being injured or abandoned themselves. They will do anything to hold on to their partners, no matter how cruel or violent, ignoring their daughters' screams and pleas, and even rationalizing that they're doing the right thing by not getting involved. They look away and silently let the harm continue, leaving their daughters feeling fearful, suspi-cious, and full of guilt, believing they've brought all this pain on themselves.

Kim: Facing Old Ghosts

Kim is a striking, auburn-haired woman of forty-two who writes for women's magazines. She told me her relationship with her daugh-ter, Melissa, who was sixteen, was starting to be full of friction and tension. Kim and Melissa had been very close, but once Melissa started the normal process of pulling away and preferring to spend more time with her friends, Kim had become preoccupied with worry. Melissa was popular with her friends and a good student, and Kim told me she wanted to be sure things stayed that way.

KIM: "She's complaining and complaining that I don't trust her, but all I'm doing is setting limits so things don't get out of control. She has a nine P.M. curfew, I have her check in from wherever she is, and of course, no dating or overnights. That's a recipe for trouble."

I told Kim that I didn't understand what she was so worried about. Melissa had good grades and seemed to be doing well.

KIM: "That's right. But I know what happens when you don't keep a close eye on kids this age. They can spin out of control in a second."

Kim seemed to be creating negative expectations of Melissa out of whole cloth, and I wasn't surprised that a sixteen-year-old would be upset about living with such binding restrictions. She couldn't even go out to an evening movie and stay to the end if she had to be home by nine. But Kim insisted that her daughter needed her protection.

KIM: "You know how bad it is out there and how easy it is for kids to get in trouble. I wish to hell that my mother had cared about me as much as I care about Melissa. There would've been a lot less turmoil in my life."

I asked Kim to give some careful thought to whether her anxiety about her daughter might be connected to issues from her own life. Were there some old ghosts dancing around?
She thought for a long moment.

KIM: "I guess I've always been worried that I would not be a good enough mother. I know that talking about this is long overdue. . . . My childhood was so awful, and I thought, 'It's

over and done with—I have a good life now. I can just grit my
teeth and go on.' But I've got so much buried garbage from
the past."

Kim's eyes filled with tears. I assured her that once we dealt
with "the garbage" head-on, it wouldn't have so much power over
her. "What was going on in your house when you were a kid?" I
asked.

KIM: "The only person I've ever trusted enough to tell about this is
   my husband. . . . My childhood was a nightmare. My father was
   a bully who had fits of crazy rage. He would beat me and throw
   me against the wall regularly. And my mother just stood by as a
   silent witness. She didn't do a thing! She allowed him to treat her
   like shit, and she allowed him to treat me the same way. I had
   to pay the price so she could have a husband and the facade of a
   family. All she cared about was what everyone would think."

In an abusive marriage, the mother becomes a terrified child—
far more concerned with defending herself against physical or emotional violence than she is about keeping her daughter safe. She
hides—sometimes using her child as a kind of shield to take the
brunt of the abuser's treatment—instead of taking the necessary
steps to get the abuser out of the house.

KIM: "I wanted so much for her to protect me and care for me. But
   she watched everything and then acted as if she was blind."

Kim became the sacrificial lamb while her mother lived in a constant state of denial. In such situations, truth becomes the enemy
because it is a threat to maintaining the unhealthy balance of a destructive family. If these mothers were to face the truth, they might
have to do something about it—call the police or a child abuse agen-

cy. But they're too frightened to even consider that. So they preach the value of silence and compliance and try to stay out of the way.

KIM: "My dad . . . was crazy. He beat me with a belt, yelled at me, punished me. I couldn't do anything right. Every day in that house was hell. I felt like I was drowning . . . there was never enough air. From the time I was five or six, I knew rage, hate, anger, and white-hot fear better than anyone should ever know it. I wanted my father to die . . . and I hated him so much I wished . . . I wished I could kill him. What child should ever have to feel like that?

"And my mother! I know she heard me scream, heard the belt hitting my skin. I know she heard the anguish when I cried for help. . . . And she never once protected me. I was her little girl and she never. . . ." (She sobbed quietly for a while, then wiped away the tears.)

"You know what I could never understand? Why we couldn't go live with my grandmother. She lived in a big house, and I always counted the extra beds and wondered why we couldn't stay with her. We had a place to go, but my mom kept me under the same roof with that monster. She let him abuse me and my little brother. . . . I told her we should all run away to Grandma's. But she told me, 'You know we can't do that. Your dad would never let me get away. Don't talk that way. It's not going to happen. Don't bring it up again.' I felt helpless and scared all the time, and I had no one to talk to. I learned my voice didn't matter—I guess that's why I tried to express myself through writing. I felt so isolated. I didn't know who I could trust."

WHEN TRUST BECOMES A CASUALTY

That atmosphere of fear, frustration, and betrayal left lasting marks on Kim's ability to read people and situations, and she couldn't develop an accurate emotional barometer. When she left home, she often went to extremes where trust was concerned; most unprotected daughters do. They may assume erroneously that everyone will hurt or betray them, and believe they are alone in a dangerous world. That can lead them to undermine closeness and intimacy by becoming fearful and suspicious, and often expecting the worst of people. After all, if you can't trust your mother, why should anyone else be different?

Or, paradoxically, they may swing to the other extreme and become overly trusting, feeling so desperate to find someone who cares for them that they may ignore warning signs and find themselves involved with people who will victimize them again. Women who were unprotected as children don't believe they are worthy of love—on an unconscious level, they believe that if they were, their mothers wouldn't have allowed them to be hurt. "I don't trust that anything good will happen for me," a woman who was an unprotected child tells herself.

"No one good and kind would really love me." Most adults who were abused as children are often unconsciously pulled toward the kinds of people and behavior that became familiar in childhood, and for daughters like Kim, that often means unstable and even potentially dangerous partners.

In college, Kim met Alex, a smart, outgoing business student, and she told me, "I felt like my life was finally going to be good. Here was someone who really seemed to love me." When he asked her to marry him a year into the relationship, she said an enthusiastic yes, even though she'd seen glimmers of his temper, which disturbed her from the beginning.

KIM: "Looking back, I can see all the little moments when I knew
     there might be trouble. He'd blow up at a waitress because
     our food took a few extra minutes to arrive. Or he'd get into a

yelling match with some crazy person in the street instead of just walking by. It made me nervous, but it didn't happen a lot, and I thought he'd just had a bad day."

Kim could see Alex's potential for explosiveness and it scared her, but she took a certain comfort in it, too—don't ever underestimate the power of the familiar. But she hadn't been destroyed by her treatment as a child, and there was a healthy part of herself left intact that could see Alex clearly.

It was that healthy part that emerged to save her a couple of years into the marriage when Alex's rage roared toward her.

KIM: "I put up with a lot from Alex. He was okay when he was sober, but he started drinking a lot, and he was a mean drunk. He had a terrible temper, and I was so scared after Melissa was born. When he got angry, he looked like my father all over again. But one night he smashed a wall with his fist and broke our best china because he didn't like what I made for dinner. When that happened, I knew I had to get a divorce to protect me and my daughter. I swore I would never be the kind of mother to her that my mother was to me."

Kim acted with considerable courage when she left Alex. Scared by how close she had come to being abused again, and how close Melissa had been to violence in their home, she found a support group for survivors of child abuse and devoured books. She found out that she wasn't alone and drew great strength from being in a community of women who understood what she'd been through.

And until recently, she believed she was finally putting the past behind her. She had done well as a writer, and her second husband, Todd, a successful chemist, was wonderful to her and Melissa. She had many satisfactions, but her painful conflicts with Melissa were troubling her deeply.

The old decision that had helped her through the hard times— "I will never be the kind of mother my mother was to me"—was standing in her way now. Kim feared that if she didn't constantly watch her own daughter, she could be guilty of turning into her mother. So she'd compensated by becoming an overprotective disciplinarian. And that old issue of trust had arisen for her again— though she knew intellectually that Melissa was responsible and levelheaded, she found herself expecting the worst of her. Once again, she didn't know how to find a reasonable center.

As we worked together, Kim started to realize how much her own childhood terrors were at the root of her anxiety about her daughter, and they both significantly diminished for her as we exorcised the pain and power of her childhood experiences. She was able to ease up on Melissa, and with time and goodwill on both their parts, they were able to reclaim the loving relationship that Kim feared they had lost.

Nina: When the Victim Becomes the Villain

Many nonprotective mothers have a shockingly well-honed ability to justify an abuser's behavior by blaming daughters for "causing" the abuse inflicted on them.

At her first session, Nina, a forty-eight-year-old computer systems analyst, told me she wanted to learn to relate better to people and improve her self-image. Short and rumpled, with her graying hair pulled back in a braid and not a trace of makeup, she'd never been in a serious relationship.

I asked her how she saw herself.

NINA (looking into her lap): "I'm homely and so clumsy. My nose is too big, my eyes are too close together. Nobody's ever going to want me. All I have to do is look in the mirror—it's no secret."

The mirror is neutral, I told her. It doesn't say words like "You're homely" and "Nobody will ever want you." But she'd heard those words on a regular basis—from her father and mother.

NINA: "I was the black sheep of the family. They wanted a pretty blond girl, and I was short and dark and awkward, always tripping over something. See, I have this really weird joint condition, and when I was a kid, it made me really clumsy. I fell all the time. My joints weren't stable, but I didn't find out the reason for a long time. My mother didn't really believe in doctors too much. She'd say, 'You do all that falling to get attention and provoke your father.'"

"To do what?" I asked her.

NINA (after a long silence): "Beat me. He started beating me when I fell. He said I was doing it on purpose. Then he would beat me up whenever he was in a bad mood. With his fists. With a strap. . . . I was afraid to fall, and I couldn't help it. When I was little, I stayed in my room until he'd left for work so he wouldn't see me."

Like so many nonprotective mothers, Nina's mother became cruel and critical, projecting a stream of blame onto Nina to justify her own cowardice and terrible neglect. "Stop making your father upset," she'd tell her terrified daughter. "Stop saying bad things about him—I don't want to hear them." She built up her abusive husband while tearing her daughter down, saying things like, "You know how hard he works—you have no compassion. You don't know how to be in a family."

Down is up and up is down in the perverse logic of abusive households. Little Nina, with her debilitating and untreated physical problems, was the villain, and her father became "the victim," even though his own child cowered and hid from him. "Just be nice to him—say good morning and smile," Nina's mother would tell her. Smile at the man who beats you.

At the same time, she'd batter her daughter's self-image.

NINA: "She would shake her head as she looked at me, like I was a curse she had to live with. And tell me how ugly I was."

With great resilience, Nina built a life for herself when she was old enough to get out of the house. She got computer training, saved her money, and moved as far away as she could. But she took her mother's words with her and replayed them in her head in an endless, self-fulfilling loop:

- You're selfish.
- You have no compassion.
- You're ugly.
- You're damaged.
- You'll never find a man.

Small wonder that Nina was so painfully shy and withdrawn. Certain that other people would hurt her, say unkind things about her, and blame her for everything that went wrong, she avoided any contact that wasn't necessary for her job, and kept to herself.

She and I began to untangle her real self from the distorted images imposed by her mother, but after a couple of one-on-one meetings, I sensed that what she needed most was a situation that could break through her isolation. Group therapy would be ideal, and since I didn't currently have any groups of my own, I referred

Nina to a trusted colleague and told her that we could phase out our work together once she felt comfortable in the group. She was petrified by the idea of talking in front of people, but after the second group session she found the courage to open up. People listened, she told me. Over time, she was able to look the group members in the eye without fear, and, for the first time, she experienced the pleasure of connecting with other people.

When Mother Is Out of Control

It's shocking to experience the betrayal of a neglectful mother. But a distinct and piercing shock comes when Mother is the abuser.

Suddenly, the hand that should be caressing curls into a fist. Or it reaches for a belt, a coat hanger, a wooden spoon. The woman whose love should be a given looks at you, or through you, nothing in her gaze but rage. And then she hits.

Her rage transforms everything. Common kitchen objects turn into weapons. A child's soft body bruises, and bones may even break. Mother becomes monster, and a world that should be safe shatters.

Early in my career, when I was working with so many adults who had been abused as children, I assumed that it was primarily the father or male figure in a household who physically abused his children. But experience has taught me that mothers do their share of hitting and beating.

These women are seriously disturbed, some even mentally ill. And when angered, they lose their ability to control their impulses. Rage takes over for the abusive mother, and her daughter is a stand-in for every person who ever hurt or disappointed her. The child triggers all her unresolved angers, resentments, feelings of inadequacy, and fears of rejection and becomes a convenient dumping ground for all the ugliness this mother has inside.

My client Deborah provides a chilling example.

DEBORAH: "Growing up, I never knew when my mom would
erupt and how mad she would get. Our home was a living
hell—constant yelling, screaming, name-calling, unpredictable
violence. She was so vicious. She slapped my face, hard, and
hit me in the head more times than I can remember. She beat
me with wire coat hangers, hitting me on my arms and hands
and back. And when I'd run into the bathroom to escape,
she'd come running after me and open the door lock with a
pencil. She screamed at me and said I was a spoiled brat and a
horrible girl. Then she'd hit me again and pull my hair. She'd
make me stand in the corner with my nose to the wall for
hours for disobeying her, and when my legs would get numb
and I'd fall, she'd yank me up by my arm and beat the backs
of my legs until I stood on my own. It was relentless . . . I can't
understand how anyone could have been so cruel to a young
child. I'm not sure how I survived."

Deborah: Learning to Deal
with Rage

I met Deborah, a forty-one-year-old graphic designer with a small
and growing business, after she e-mailed me asking for the earliest
appointment I could give her. She'd had a blowup with her eight-
year-old daughter and was terrified by the anger she felt. "I'm in
trouble," she wrote. She was pale and anxious when she came into
my office a few days later, and after I'd gotten a little background
information from her, I asked her to tell me what was going on.

DEBORAH: "I almost hit my daughter the other day and it really
scared me. I was so angry I couldn't see straight, and I think
I could've hit her. I didn't, but I was this close, and that's one

thing I've always sworn I would never do. . . . It's no excuse,
but I've been under so much pressure lately. We've got three
kids under ten, and my business is growing, which is great,
but I'm working all the time and I'm wiped out when I get
home. I walked in Thursday night and Jessica, my eight-
year-old, was in the living room by herself watching TV—
the other kids were with their dad upstairs watching a ball
game. I don't know what got into that girl. She'd made a fort
out of the sofa cushions and dragged a bunch of food in. She
must've been roughhousing with the dog because there was
popcorn scattered all over the place and a stain where Coke
had spilled on the rug. And she was just sitting there in the
middle of it, watching some inane show. I grabbed the remote
and turned off the TV and laid down the law. I told her to
clean up the mess, go up to bed as soon as she was done, no
TV for at least a week, and no snacks in front of the TV till I
say so.

"She just sat there. And when I told her to step to it, I could
hear her calling me a mean old hag under her breath. I just
snapped and started screaming at her. . . . It was awful. 'How
dare you talk to me like that. Who the hell do you think you
are, you ungrateful little bitch? I've had it with you. I work my
ass off for you. . . .' I never, ever talk to the kids like that. The
dog's leash was on the table and I reached for it and I felt my
hand go up like I was going to. . . . Oh my God, Susan. Jessica
was terrified. I knew that look so well. That was me when I was
a kid and my mom was about to hit me. Am I turning into my
mother? . . . I can't let that happen. My mom was crazy. Am I
crazy, too? I seem to have so much anger."

I reassured Deborah that anger is just a very strong feeling. It
doesn't mean you're crazy. Deborah had every right to be upset,
but as she had learned, screaming and hitting don't teach a child

anything positive. Rage only teaches rage. Deborah would need to work on the anger she had pent up inside. And to do that, we'd have to look closely at the abuse she'd experienced as a child.

Deborah told me that she'd been beaten by her mother from the time she was three or four, and in vivid detail, she described the terrible forms that battering took. When she was old enough to leave home, she said, her mission was to put the violence behind her and never let it back into her life. She cut off all contact with her mother when she went to college, even though that meant working a couple of jobs to support herself. One of them, with a graphic design firm, led to a full-time job when she graduated, and a few years ago, she left to open her own boutique Web design company.

DEBORAH: "I really thought everything would be okay when I stopped being in touch with my mom. Especially once the kids were born and I had my own family. Once you've had a child of your own, it's hard to imagine how anyone, especially your own mother, could hurt her own little girl. This was the woman who was attached to me with an umbilical cord. I was inside her body. I know what it feels like to sense a baby growing in your womb, to see its face for the first time. . . . And to be so savage. . . . How could she? It makes me so livid to think about it."

Deborah, like so many abused daughters, had a volcano of rage inside because of the pain, humiliation, and degradation she had suffered at her mother's hand. And now, having seen it explode toward her own child, she lived in terror that it might spill out of her again. It was a legitimate fear: Without treatment, the intense emotions surrounding physical abuse can make daughters vulnerable to becoming abusers themselves.

IT'S OKAY TO SAY "I'M SORRY"

Deborah knew that her first priority was to calm her relationship with her daughter. "Jessica is practically hiding from me," she said. "She's still scared, and I don't know what to do. I think I really traumatized her."

I suggested that she start with an apology. Apologizing when you have been wrong is a great gift you can give your child. It lets her know that you are not afraid to be vulnerable or honest, and that you respect her enough to acknowledge your mistakes. It was also appropriate to ask for improved behavior on Jessica's part. I told her, "You need to ask her to respect you enough to know how hard you work and that you're tired when you come home—and that you really need her to clean up whatever mess she makes."

The apology went well, Deborah reported back. When she reached her arms out to Jessica afterward, her daughter came for a hug and melted into her as Deborah stroked her hair. Now Deborah was intent on rooting out her rage, and we spent our next sessions focusing on her anger—and her grief.

The Double Betrayal
of Sexual Abuse

Daughters pay an unfathomable price when their mothers are aware that they are being sexually abused and do nothing. Sexual abuse shrouds a daughter in deep and pervasive shame that leaves her feeling fundamentally violated, stigmatized, and alone. She sees herself as "damaged goods."

Even after years of candid talk about the subject, many people are still unclear about the driving force behind this crime. That force is not primarily sexual, but a cold, life-warping need for power and control on the part of the abuser, who uses his authority to get his victim or victims (as he may molest more than one daughter) to comply. He may also manipulate or cajole—"Make Daddy happy," "Let me show you what to expect when you start seeing boys"— making the daughter feel complicit in the abuse, and saddling her with ever deeper layers of guilt and shame that rightfully belong only to him.

The predator wants what he wants and he takes it—from an innocent and powerless girl who may be three years old or seven or in her early teens. Even if he's aware on some level that she will be severely traumatized by this violation of her body and her essence, and crushed by a trusted adult's betrayal (and it's difficult to believe that he doesn't have *some* awareness), none of that gives him pause. Emotionally infantile and insecure, a sexual abuser is deeply dysfunctional and severely disturbed in his personal life, no matter how well he may function in the outside world.

And what about the mother who may know or suspect what's happening but continues to pretend that everything is fine? Like the other mothers we've seen in this chapter, she's excessively dependent, afraid to challenge the abuser whether he's her husband, boyfriend, or another member of her family—and unwilling and unable to pull her daughter to safety.

Sexual abuse only occurs in deeply troubled families where role definitions and boundaries are totally blurred and violated. I have treated a vast number of victims and guided them through their brave journey to regain their confidence, their dignity, and most of all, their self-respect. For this section, I have chosen one representative case that offers a window into the collusion of the silent mother and the abuser. If you were sexually abused and unprotected, I think you'll find many elements of your experience here. And I want to assure you: You, too, can heal. The process starts now, with fearlessly facing what happened.

Kathy: Wounds You Must Tend To

Kathy was a smartly dressed thirty-three-year-old who worked as an account executive with an ad agency. She told me she was concerned that her two young girls were suffering because of her recurrent depression, which she wisely recognized as being triggered by the long-term effects of untreated sexual abuse by her father. Her story was all too familiar.

KATHY: "I've been struggling with this for most of my life. My father started abusing me when I was eight years old. It was horrible. . . . I tried to tell myself it could have been worse, that other people have suffered so much more, but since I've had kids, I've noticed how the memories have gotten so much stronger. Anyway, I get really sad, and I'm here because I don't want my babies to think they're the cause of my pain. I noticed that my older daughter gets tummy aches when I'm in the dumps, like she can sense my depression. She doesn't deserve that. So I think it's time to see if I can really deal with my past. I've done a lot of reading and tried to do some work on myself over the years. I thought I was better, but I was wrong. I'm not done."

Kathy was wise to come in. Sexual abuse is one life experience that absolutely mandates professional help. Depression is as constant as the change of seasons for people who've been through what Kathy had experienced. But the more you work with a good therapist, the more the memories of abuse lose their power over you. Doing that work is a loving gift to yourself and your family.

The first step, I told Kathy, was to talk about what had happened in her house when the abuse was taking place. Naturally, it was difficult, but she summoned her courage and plunged ahead.

KATHY: "The bad stuff started when I was eight. We'd be sitting
on my parents' bed watching TV and my dad started wanting
to play 'ride the horse,' with me bouncing against what I
learned later was his erection. I didn't know what was going on
at first. Then he started to put his hands and his mouth on me,
and he made me touch him. . . . He never penetrated me. But
it was awful, Susan. . . ."

SUSAN: "Of course it was. You were confused. You were
frightened. And you don't have to have been penetrated to be
sexually abused."

Sexual abuse encompasses a whole range of actions that may
or may not involve penetration. All of them involve the betrayal of
trust and the wielding of the abuser's power to coerce or involve
the victim. Exposing genitals to a child, showing her pornography,
and asking her to undress and expose herself to him fall on this
spectrum, even though there may not be actual physical contact.
When there *is* contact, abuse can take myriad forms—touching the
child's genitals, buttocks, or breasts or having the child touch the
adult's; rubbing against the child; penetration with fingers or objects; oral sex; intercourse.

Your body—all of it—is where you live, and your whole being
feels the impact when it is violated. Bottom line: Any kind of behavior with a child that has to be kept secret probably falls under
the heading of sexual abuse. And like all the examples given above,
it's almost certain to be a criminal act.

The Silent Partner: Denial and Accusations

Kathy's abuse went on for years, and I asked her if she'd ever told
anyone.

KATHY: "My father warned me not to say anything, but I told my mother when I was ten. I wanted it to stop! But she essentially did nothing! She talked to my dad, and he said he would never do it again, that he would get counseling. None of it was true. The sexual abuse went on and on."

A loving mother, knowing that her daughter was being molested, would rear up in fury and take steps to end the abuse. "If anyone touched my baby like that," one caller to my former radio program told me, "I'd want to kill him, and I'd call the police in a minute!" She was the epitome of the protective warrior mother, and every daughter deserves a mother like that. But a daughter whose mother lacks that righteous anger and strength may be abandoned for years to attacks on her body and being.

Worse, her inadequate mother may make her feel responsible for her own abuse, as we saw earlier with Nina. The words are as corrosive as acid when such a mother blames her victim daughter:

- He would never do something like that. You must have come on to him.
- You could have stopped him if you wanted to.
- You must have enjoyed it.
- If you hadn't worn those tight shorts, this never would have happened.

She may flat out deny that the abuse is happening with words like: "You're making this up to get attention." "That's impossible." "You're saying this to get back at him."

If she deigns to "protect" her daughter, it often takes the form of an ineffectual "Put a lock on your door" or "Just stay away from him."

How is this kind of denial, callousness, and complicity possible?

Like the other mothers in this chapter, the woman who allows her daughter to be sexually abused is passive, fearful, self-absorbed. She may be terrified of what might happen if the family were split up. She may be afraid of the shame or guilt she'd feel if others were to find out. She may believe that her daughter's abuse is the price she has to pay for her husband's financial support of the family and fear the turmoil and consequences she would have to face if she took action.

In some instances, she may even be jealous of her daughter. It's not uncommon for such a mother to feel that she's been replaced in the marriage, mistaking the brute power dynamic of sexual abuse for a sexually based act, and seeing her young daughter as a competitor for her husband. If the father is a successful professional, as many incest perpetrators are, she may not want to give up the goodies that go with that—financial security and a big house are often more important to her than her daughter.

This crippled mother is almost completely devoid of empathy and compassion. Love and protection aren't in her emotional vocabulary.

THE SECOND LEVEL OF BETRAYAL

I can't overstate the impact a mother's response has on the way a daughter heals from any kind of abuse. It's crucial to the way her daughter thinks about what's happened to her and the way she feels about herself in the aftermath. A loving mother believes what her daughter tells her, assures her she did nothing wrong, and takes action to ensure that the abuse will never be repeated, often by getting a divorce or having the abuser arrested. In the absence of this essential validation, the abused daughter feels damaged, dirty, and different—the Three *D*'s of incest.

Kathy coped initially by isolating herself, as some victims do, hiding behind a wall of weight, with the erroneous belief that *that* would make her less desirable and therefore safe.

KATHY: "I had no interest in dating for a long time. Who would want me? I was the girl whose own father did awful things to her. I ate to fill the gap, the loneliness. I didn't trust anyone, and I was stressed all the time. In college I put on a lot of weight, and that made me feel even worse about myself. I got some counseling for my depression and managed to lose a lot of it, but I was still convinced I would never feel loved. . . .

"After college I got an internship at an ad agency, and then a real miracle happened."

At work, Kathy began a friendship with Ethan, a man who was kind and playful. Their mutual attraction became a romance.

KATHY: "Ethan has really been great. I know it hurts him to know the specifics of my pain. He's heard it for the thirteen years of our relationship, and he's stood by me while I've been trying to get better. He's been a godsend."

But even with Ethan's love and support, she was never sure when memories of her abuse, which were largely dormant early in the relationship, would overtake her. They flared up when each of her daughters was born, and sometimes when her husband bathed or dressed the girls. This is not uncommon; having children is one of the most powerful triggers for reactivating dark memories. Other triggers can include the death of a parent, a scene of abuse in a television show or movie, even seeing your daughter reach the age you were when you first were abused.

KATHY: "Mom feels that this should be behind us, and she recently told me that she won't talk about it because it is embarrassing to her. She has no idea what embarrassment is. I'm at a point where I don't want her denial and negativity in my life. She acts as if nothing happened. I want to put this behind me, too, but she won't help. And I am so furious about that. They say you need to forgive before you can move on. I wish I could."

SUSAN: "What your mother did was terrible, Kathy, and there's no need to forgive her despite what a lot of people may tell you. But you *do* have to release the power and control that her betrayal has had over you. Forgiveness is not a magic wand that you can wave to change everything, especially when your parents have done nothing to take responsibility for their destructive behavior."

KATHY: "Thank you for saying that. Now that I have my two beautiful babies, the outrage keeps surfacing. I could never let anyone hurt my children. I would never put them in a situation where anyone could *potentially* harm them. I guess my challenge is to figure out why my mother didn't feel the same way about me. . . ."

I told Kathy that it's not usually productive to focus on the "why," because we may never get to the bottom of it. Healing comes from looking at what happened, how it affected you, and what we can do about it now.

Wounded, Not Ruined

Of all the things I've done in my long career, I'm proudest of the fact that I was one of the very first mental health professionals to take sexual abuse out of the cave of secrecy where it had been hidden and bring it into the light. It was a tough battle, and I talked (oh God, how I talked) on radio, on television, in seminars, and in newspaper interviews, until the public—and some very resistant members of the psychiatric profession—were willing to listen and sexual abuse became a subject that was no longer taboo. Today there's a much greater understanding about the prevalence of such abuse, and the depths of the wounds it creates. There's a better understanding, too, of physical abuse and neglect.

I am always deeply moved and awed by the courage and determination of women like Emily, Kim, Nina, Deborah, and Kathy. Despite the betrayal of their mothers, with treatment, they not only survived but went on to lead fulfilling lives.

I want to reassure you that if you are a victim of emotional neglect or physical or sexual abuse and were not comforted or cared for appropriately, your life is not a dead-end street. As severe as are the effects of neglect and abuse, the healing can be equally dramatic. You are not doomed, or cursed. You are wounded, not ruined, and great wisdom can come from your wounds: compassion, empathy, a sensitive barometer that lets you know when people are mistreating you. Using that wisdom is how we all make lemonade out of the lemons. Let me show you how.

Part Two

❦

Healing the Mother Wound

In the chapters that follow, you'll see the highly effective techniques—role-playing, letter writing, visualizations, and powerful exercises—I used to take many of the women you have met in this book from their bewilderment, hurt, and anger to the freeing realization that the pain and confusion they felt with their mothers wasn't their fault.

As you follow along, and do some of the exercises yourself, you, too, will come to know that you are not responsible for your mother's unloving behavior. There was nothing about you that deserved to be emotionally abandoned or smothered or abused or bullied or used to build your mother up while grinding you down. *You deserved to be loved*, and your mother didn't or couldn't give you the warmth, safety, support, and feeling of being cherished that you needed so much. Some of our work in these chapters will be aimed at helping you accept that truth at the deepest levels of your being. Before you can heal—and make sound decisions about how to handle your relationship with your mother—your head, your gut, and your heart must all fully acknowledge what happened to you.

I know from personal experience, as well as from my clients and the letters and e-mails I receive, how much that acknowledgment hurts. But once you have the courage to accept the truth, you can free yourself from the wounds of inadequate mothering. Go at your own pace. Read a little of this book and then take the time to let what you've read sink in before going on. Don't try to stuff your feelings. You will probably cry a lot—and that's fine. Grief and anger are a natural, necessary part of this process. Remember, tears are like rivers that start in one place and flow to another—they can help carry you to healing.

Please keep in mind that the exercises in these chapters are not intended to replace in-person work with therapists, support groups, or 12-step programs. If you experienced physical or sexual abuse growing up, it is *imperative* that you get professional help. If you are currently using drugs or alcohol to ease your pain, you need to join a 12-step program or work with an addiction specialist and be sure that you have at least three months of sobriety before taking on the work in this book. In the early stages of recovery, you are extremely vulnerable, and your emotions are very raw. Delving into difficult memories of your childhood at that point could cause you to slip back into old patterns of substance abuse. Also, if you are or have been struggling with depression, you need to know there are many ways of treating it now. It's vital to know that you are not alone.

We'll begin in the cool, cognitive realm of the head before getting to the work we'll do with the heart. A healthy person is able to think *and* feel, and I'll help you balance both. If you don't fall into the categories I described above, you certainly can do some of this work on your own; the communication exercises in particular are safe for most anyone, and they're highly transformative. But remember that you'll get the greatest benefit by using this part of the book in conjunction with professional support.

You will be dealing with feelings that can be overwhelming, and it's important that you have someone to turn to as you embark on this journey. If you're in therapy or plan to be, you may want to take this book with you to your sessions and do some of this work in that setting. It's a sign of strength, not weakness, to seek out the help you need. Remember, too, that you can return to these pages again and again for guidance, validation, and support.

Some Guidelines for Choosing a Good Therapist

If you decide that you'd like the help of a therapist, be sure that the person you're going to work with is comfortable and experienced in dealing with unhealthy families and the damage they do. If your relationship with your mother has been painful and scarring, you need a therapist who isn't afraid of diving into the muck with you so that you can come out stronger and healthier.

Please don't stay with a therapist who hears your history and says things like:

- That's all in the past—you need to move on.
- Let's just deal with the here and now.
- You need to cut your mother some slack. She had problems, too.
- You don't want to spend your life feeling sorry for yourself.
- You need to forgive and forget and get on with your life.

All of these comments are dismissive, and they discount your feelings and experience. Working with a person who approaches your past that way will only confuse and frustrate you, further re-inforcing the self-blame you may already feel. ("Why am I being such a baby?" "Why can't I get over this?") Look for a therapist who works actively with you instead of just sitting back and saying, "Uh-huh" or "How do you feel about that?" You want a person who gives you feedback and actively engages with you. Trust your instincts. If you don't feel comfortable, safe, or truly heard when you're with a therapist, he or she is not the person for you.

Be Gentle–There's No Rush

Be very kind to yourself, and during the time you're using this part of the book, give yourself extra time to write, walk, think, and rest. Your emotions may be volatile and heightened as you work with the material in some of these chapters, so don't make any major decisions until you are calmer and able to think clearly and rationally. If you are having problems in a love relationship and are confused about whether you want to salvage it, don't make any impulsive moves until you have worked on the situation with your mother.

I can't promise you that simply reading these chapters will magically change your life and heal the damage that came from your mother's unloving behavior—that would be irresponsible of me. What I can promise is that if you use the strategies I will teach you, your pain and confusion will lessen. You will be able to see both yourself and your mother in a new, true light. That vital clarity will provide a platform from which you can make healthy decisions about your relationship with her, and rebuild your life.

Chapter 7

The Beginnings of Truth

ᴡ

"I'm starting to see it wasn't all my fault."

- You say yes when you want to say no.
- You vow to stand up for yourself, but you retreat again and again.
- You can't find a way to evolve, claim your own life, and step out of your mother's shadow.

None of this is logical. Rationally, you know you have many choices. "I'm an adult. I should be able to tell my mother I can't have lunch with her without being swamped with guilt," you tell yourself. "I'm able to reschedule lunches with friends—so what's the problem? Why is it so hard to do something so simple?"

The answers lie in your programming, the messages from your mother that parachuted into your being like a thousand dandelion seeds, planting in you false beliefs about yourself and your relationship with her. In a sound mother-daughter relationship, the messages you received would have been full of nurturing, building your confidence and supporting your growth and your moves toward independence.

But for the most part, that didn't happen. Your mother was focused on her own needs far more than on yours, and she was often consumed with her attempts to deal with her own distress. We've seen how common it is for an unloving mother to rationalize and deny her wounding behavior and deflect the blame onto you. We've also seen how common it is for daughters to accept that blame and to fall into self-defeating patterns that have taken hold in their relationships with their mothers in their adult lives. That's what happened to you. You were programmed with a flawed set of messages that predisposed you to work against your own best interests and put your mother's priorities first.

Those messages have been delivered not just in words, but also in your mother's behavior toward you and even in her body language: sighs, eyes rolling in disapproval, smiles when you comply, angry silences when you don't. This constant instruction and feedback from her, so much of it devoted to keeping the balance of power in your relationship tipped in her favor, has warped and limited your basic sense of your identity, your worth, your goodness, and your place in the world. Even if you are distant from your mother today, much of your life has been shaped by the early programming you got from her. Before you can shift it, change your relationship with her, and reclaim a full and true sense of who you are and what's possible for you, you'll need to uncover the untruths embedded in the messages she sent you, then, step by step, begin to take that self-defeating programming apart.

That's what we'll begin to do in this chapter.

This work is powerful, and it's demanding. We'll move slowly, focusing first on how the programming process works, and then looking closely at the factor that is the easiest to access—your beliefs.

Programming 101: "You Are" Becomes "I Am"

A mother sees her toddler struggling to walk, smiles, offers a hand for support and says, "You're amazing! Look at you, you're *walk-*

ing. You're a little gymnast already." In that moment and thousands more like it, messages fly toward the child, who takes in everything: "Mom notices and cares about what I do. She loves me. I'm amazing. I'm walking."

The smiles and good opinion of her all-powerful mother mean everything to the dependent daughter, who strives to do what she can to prompt more. Harshness and criticism, by contrast, can be terrifying: "If I displease Mom, I may not survive—she may leave me to fend for myself," the child believes. But whether the messages are positive or negative, the child absorbs them and builds a core understanding of herself around them. The mother's "you are" becomes the daughter's "I am."

These internalized messages help form some of our oldest and deepest beliefs: Because they've been part of the air we breathe for so long, we regard them as true and often behave as though they are true without ever questioning them. That's a wonderful thing when we've been praised and encouraged since childhood; it allows us to emerge with beliefs that tell us "I'm strong and capable," "I'm a good person," "I'm resilient," "I'm lovable." But unloving mothers have loaded their daughters up with beliefs—deeply ingrained concepts, attitudes, expectations, and perceptions about ourselves, the people in our lives, and what's right and wrong—that are false and highly destructive.

Many of your mother's "you are" statements have reflected her disapproval, criticism, or helplessness: "You are so selfish." "You are the only one who can take care of everything." "You are upsetting me and making me feel ill." When these messages become internalized as beliefs, they don't just sit inert inside you. They trigger painful feelings. You fight the false beliefs that tell you what a bad, thoughtless, selfish, or incompetent person you are. You argue with them, wondering if they're true and what it would mean if they were. You try to prove them wrong. Mostly, though, you suffer. You feel sad, angry, guilty, embarrassed, burdened, ashamed, bitter, defiant, bad, resigned—the emotional possibilities are many, and all of them hurt.

What happens to the painful feelings? They drive self-defeating behavior. If your narcissistic or controlling mother has programmed you with the false belief that you'll never measure up to her (or anyone's) standards, you may feel insecure, inferior, inadequate, lacking in confidence. And because of those feelings, you may well hold yourself back from success in relationships by aiming low, or shrink away from the career you want by telling yourself that since you didn't get the first good job you applied for, you'll stop trying. Why set yourself up for more shame or disappointment? After all, a voice in your head tells you, "I'll never be good enough. I couldn't possibly compete."

Whose voice is it? Your mother's. Whose interests does it serve? Hers. The narcissistic mother doesn't have to tell you "Push yourself down so I'll look better," and the controlling mother needn't say, "Prove me right by failing." The programming she's instilled does the work even when she's not there.

When you find yourself locked in behavior that seems self-defeating, you can be sure that this cycle is at work: A false belief is creating a painful feeling, and you are choosing unhealthy behavior without realizing it to avoid or ease the pain.

Let me give you a more detailed example of how the cycle works.

I. As a child, you take in your mother's messages.

From the time you're little, your depressed mother tells you, "I can't make it without you. You're the one who's holding the family together. You're my little angel for doing all this." Her only smiles come when you've cooked dinner for the family (at age eight) or called her boss to say she can't come in because she's sick when she's holed up in her room watching TV.

2. YOU TRANSLATE THOSE EARLY MESSAGES INTO AN ARRAY OF FALSE BELIEFS.

"I'm the only one who can make my mother happy. I have to earn her love with 'good deeds' that make her feel better, even if that means lying for her. If she's not happy it's my fault. I have no right to do what I want to do, and no right to complain. My job is to take care of her."

These beliefs, which have to do with your rights, responsibilities, and identity, as well as your mother's, set you up for the impossible. The truth is, you *aren't* responsible for your mother's happiness, and you can't fix her; only she can do that. You will always fail. Real love isn't something a child has to "buy" with good deeds. You're entitled to a childhood, and a life of your own, and it's unreasonable to think that you can give them up, much less do it without complaint. Your real job as a person is to individuate and build a life of your own, and your mother's role is to help you. But if your beliefs echo your mother's messages, not any sort of rational truth, throughout your life they color your feelings and behavior.

3. YOUR FALSE BELIEFS LEAD TO PAINFUL FEELINGS.

Confronted with your inevitable failure to fix your mother, you may feel inadequate, guilty, flawed, bad, and ashamed, both in childhood and as an adult. You were *supposed* to be able to do it, according to your programming. You may also feel burdened, resentful, ashamed of your resentment, and terribly sad about what you've missed while trying to hold things together for your family.

4. To ease those painful feelings, you turn to self-defeating behavior.

Trying to assuage that array of painful feelings can take you in many directions. As a child, it's likely that you felt driven to devote huge amounts of your time trying again and again to repair the many places that felt broken in your mother's life. *That* would prove you were a good daughter, worthy of your mother's love. You may also have mastered the art of pretending things are fine when you're struggling, and not asking for help when you need it because you believe it would reveal your supposed weaknesses, flaws, or inadequacies.

It's highly likely that as an adult, you will still be hyperaware of your mother's needs and jump to meet them, even if you don't want to—even when you know intellectually that it's not necessary or even helpful for you to try. Because the fastest, most familiar way to prove to yourself, your mother, and the world that you're competent, good, and not deeply flawed is to cater to her needs.

Human behavior is complex, and I don't want to suggest that every daughter of a depressed mother experiences this precise chain of false beliefs, painful feelings, and behavior that's not in her interests. Every mother is different, and so is every daughter. But I can say with certainty that if you trace your self-defeating behavior back to its origins, you'll find layers of programming in the form of negative beliefs and the feeings they create.

The Power of Beliefs and Feelings We Can't See

Once you understand the connection between beliefs, feelings, and behavior, it would seem fairly simple to interrupt the cycle. But even though challenging false beliefs is, indeed, a pivotal step toward making significant changes in the way you respond to your mother's needs and expectations, the actual doing is made difficult by a couple of factors. First, the nature of beliefs is that we assume they are true: The earth is round, the sky is blue, and my job as a daughter is to make my mother happy, regardless of my preferences. Many of our false beliefs have been our "truths" for so long that we don't think to question them. They become our reality, and we don't see them as filters that color our perceptions.

Quite often, we can't even identify the beliefs that give us the most trouble, nor the full range of painful feelings that are propelling our behavior. They're well hidden in our unconscious, the mind's vast storehouse of urges, emotions, thoughts, drives, fears, memories, and experiences that exist, and influence us, without our awareness. Generally, the material in the unconscious is so uncomfortable for us that it has been pushed out of sight to protect us from our deepest shames, insecurities, and fears about ourselves.

As we pull back some of the curtains that hide so much of our inner world, we begin to see just how much power our unconscious mind has. Even when a daughter consciously tries to set her own agendas and navigate the present, her unconscious is often working frantically to salve the wounds of the past. Again and again, it tries to devise ways to "make Mother love me," not only by playing out her programming but also by seeking situations in which it can set right childhood failures by replaying similar scenarios in search of a better ending.

In that way, unconscious programming often picks out our husbands, determines how much success we're allowed to have, and shapes the quality of our relationships and emotional well-being.

- **You may sabotage yourself in love:**

  CONSCIOUS BELIEF: I'd love to have a wonderful partner.

  UNCONSCIOUS BELIEFS AND FEELINGS: I'm not entitled to love or attention. I can't compete. Who would want me? I can't bring home a smart, successful, loving guy—Mom will flirt with him, or rip him to shreds. I can't be happier than she is. I don't deserve that.

SELF-DEFEATING ACTIONS: You mistrust overtures from interested people, choose incompatible or inappropriate partners. You become a rescuer and a caretaker for people who won't take responsibility for themselves. You rule out the best candidates, reasoning, "I'm a realist. I'm not going to set myself up for disappointment."

- **You may sabotage yourself at work:**

CONSCIOUS BELIEF: I really want to be successful.

UNCONSCIOUS BELIEFS AND FEELINGS: I'm not allowed to outshine my mother. I'll never amount to anything. I will find ways to sabotage myself so I can fulfill my mother's negative expectations of me. She knows who I really am. I'll never really measure up. I'm inadequate.

SELF-DEFEATING ACTIONS: You show up late, leave part of your work undone, pick fights with your coworkers, procrastinate, miss important deadlines, don't follow through on leads and ideas.

- **You may sabotage your deepest desires:**

CONSCIOUS BELIEF: I love making other people happy. I put them first because I love the feeling of being a giving, caring person. We all have to look out for one another.

UNCONSCIOUS BELIEFS AND FEELINGS: If I give up what I want, and do things for other people, I'll win their love and approval. If I can get enough love, approval, and admiration, it will make up for how bad I feel about myself.

SELF-DEFEATING ACTIONS: You paste on a smile and stuff your resentment at being taken for granted. You say "I don't know" or "I don't care" when asked your preferences. You avoid con-flict at all costs. You forget you have dreams of your own.

Clearly, if the status quo is going to change, it's vital to bring your unconscious programming to light.

What Do You Really Believe?

To uncover the messages you heard from your mother, which you unknowingly adopted as your truths, we'll have to work backward from what we can see. So let's take a look at some of the more typical messages you may have heard from your mother or intuited from her behavior toward you. Remember that your mother may not have put all these messages into words. She may simply have behaved in a way that let you know when she was upset with you, and she may continue that behavior today. For instance, she may pout or shoot you an angry look anytime you displease her. Such nonverbal messages have all the power of verbal ones, so as you go through the lists below, think about your mother's manner as well, and the way it reinforces her words.

Put a check next to the items that resonate for you, and feel free to add any that aren't listed. If you replay in your mind the expressions, criticisms, demands, and scenarios that commonly unfold as your mother gets her way, you should be able to fill in any gaps.

False messages that demean you:

• You are so selfish.
• You are so ungrateful.
• There's something wrong with you.
• You can't do anything right.

- You don't know how to be loving.
- You only think about yourself.
- You are such a disappointment to me.
- You'll never amount to anything.
- You're the reason I have so many problems.
- You'll never find a man.
- You'll never be as attractive, smart, accomplished, or desirable as I am.
- You have terrible judgment.
- No one cares what you think.
- You're nothing but a burden.
- You're more trouble than you're worth.
- You're the cause of all the trouble/abuse/shame in the family.
- If you were a better person, the abuse/trouble/shame never would have happened.

These are the sorts of words that often come from narcissistic, competitive, controlling, or abusive mothers. Statements like this tear you down, making your mother feel all-powerful while absolving her of any responsibility for her life and her discontents. If a pang of familiarity hits you in the stomach or you hear your mother's voice in your head when you read a particular item, it's likely that it's circulating inside you like an indictment. Recognizing these messages is an important first step toward taking their power away.

False messages that unfairly burden you:

- You are my whole life.
- You are the best part of me.
- I don't need anybody except you.
- You're the only one who cares about me.

- You're the only one who can keep the family together.
- We're so close we have to share everything; no secrets.
- You're my best friend.
- You'll always be my little girl.
- You're the only one I can count on.
- I need you so much—I couldn't make it without you.
- I love you so much more than I care about your father.
- You have to help me figure out what to do with the rest of my life.

This second group of messages is different but no less destructive. These are the messages that place the burden for your mother's and the rest of the family's well-being squarely on your shoulders.

They can seem seductive on the surface, but they have a desperation and a smothering quality that's palpable. They often emanate from overly enmeshed mothers and mothers who, through their inadequacies, put you into a role reversal with them.

False messages from your mother about your role and what you owe her:

- It's your responsibility to make me happy.
- My feelings are more important than yours.
- It's your job to earn my love.
- It's your job to take care of me.
- It's your job to obey me.
- It's your job to respect me, and that means doing things my way.
- Honor thy mother means you should never get upset with me.
- You have no right to challenge me or say anything bad about me—after all, I gave you life.
- You have no right to disagree.

- It's your job to stay silent if I betray you.
- It's your job to keep peace in the family by not rocking the boat or resisting what I want.
- It's your job to protect the family secrets.

As well as teaching you who you are as a person, your mother also taught you about your role as a daughter, letting you know what she expected and defining who you were supposed to be in terms of her needs. It wasn't in your mother's interests to teach you that as you matured, you were supposed to take charge of your own life. Her programming emphasized your duty to her and usually gave scant recognition to your duty to yourself.

The templates for large portions of our lives come from the beliefs instilled by the messages on these lists. They shape what is permissible, how much you are allowed to have, and which choices you'll punish yourself for making. As you look through the items you've checked and added, you're seeing reflected back to you a picture of what you believe about yourself.

You'd think that as an adult, you could stand back and say, "Sure, my mom said those things, but I know they're not true and they don't affect me anymore." But if you have never taken active steps to challenge those messages, and you're struggling through a painful relationship with your mother, those very false beliefs are almost certainly still running the show.

I want to caution you about one final set of false beliefs that's particularly problematic for daughters:

- If only my mother would change I would feel better about myself.
- If only she'd realize how much she hurts me, she would be nicer to me.

- Even though she can be pretty mean to me, I know she has my best interests at heart. I'm overreacting.

These "if only" beliefs keep you stuck in an alternative reality of yearning and longing. They keep you passive and reactive rather than proactive because you are waiting for your mother to change instead of doing the tough work of changing yourself.

It's time to stop waiting. It's time to reclaim your own power.

SEPARATING LIES FROM TRUTH

There's another word for the false beliefs that have been running your life: *lies*. And now, I want you to call them by their correct name and feel the enormous gratification of telling yourself the truth about who you are.

With the Lies and Truth Exercise below, you'll actively challenge your false beliefs. The exercise will convey your truths powerfully to both your conscious and unconscious mind. It's designed to reinforce your dignity, self-respect, and confidence, and I know you'll find it both wonderfully clarifying and freeing.

LIES AND TRUTHS EXERCISE

Part One

Take out a sheet of paper and draw a line down the middle. At the top of the left-hand column, write in bold letters the word LIES. At the top of the right-hand column, write TRUTHS.

Now, in the lies column, write the lies you can remember being told about you by your mother, the ones that really wounded you. Put the lies down as they were said to you:

"You are a . . ." (You might want to scan the lists of false messages above to be sure your list is complete.) Next to each lie, write the contradicting truth in the truths column. The best way to challenge the lies is to give specific evidence that they're wrong. Stand up for yourself. Your truth is valid— much more than the distortions your mother has fed you. Even if you don't completely believe the words in your truths column right now, they will light the way for you and show you the person you want to be and are becoming.

Many people do this exercise with great ease and relish, but some women have difficulty counteracting opinions they've heard all their lives. If you get stuck, pretend you're talking to your best friend, or a loved one who has described herself to you with the false beliefs in your lies column. What would you say to her? How would you challenge her narrow, hurtful description of herself? What would you say if someone said things like this to your daughter? It's often easier for us to defend other people, and see the good in them, than to stand up for ourselves.

Let me give you some examples of what my clients have written.

LIES	TRUTHS
You're selfish.	I'm generous, giving, and considerate of others.
You're so unforgiving.	I'm forgiving if people take responsibility for their behavior and make amends.
You're too sensitive.	I have feelings and vulnerabilities that make me a more open and loving person.

You owe me respect.	In a healthy relationship, respect is a two-way street. I respect my conscience and my integrity.
You're not a good daughter.	Anyone would be proud and happy to have a daughter like me.
It's your job to make me happy.	I've tried my damnedest but nothing is ever enough for you, so I'm quitting this lousy job.
You can't survive without me.	Just watch me.
I can't survive without you.	You're going to have to find a way. I'm no longer willing to be controlled by guilt and obligation.
You should put my feelings first.	I did that for a long time, but now I have a family of my own, and they come first.
You can't do anything right.	You're jealous of me and can't stand it when my life is going well.
You're not good enough to succeed.	I will succeed despite you and your attempts to tear me down.
You have to take care of me.	Where is that written?
It's your fault I drink/take pills/ I'm so depressed.	I don't accept the blame for your destructive behavior. You need to get help.
You will always be my little girl.	I'm an adult with a life of my own. I choose freedom, not smothering.
No one will ever love you the way I do.	I sure hope not.

When you make your list, I suggest you put about ten items in each column. Your truths can be as long as you want to make them.

Part Two

When you've finished your list, cut the column of lies from the paper on which you've written it, crumple up the lies, and, in a safe place, burn them. As you do this, say out loud: "I am now burning many of the lies my mother told me about myself. I am now burning many of the false beliefs I had about myself. I am now reclaiming the truth and my good feelings about myself."

Take the ashes and dispose of them somewhere outside of your living space—don't put them in a wastebasket or flush them down the toilet. These ashes have powerful negative energy in them, and you want them gone. Bury them in a vacant lot or put them in a Dumpster on the street. Get them out of the place where you live.

Part Three

Now (and this is the fun part), I want you to go to a party store and buy a helium-filled balloon. Take your list of truths, fold it into small sections, each containing several truths, and attach your truths to the string of the balloon. Then take the balloon to a place where you feel good—the beach, a lake, a lovely park, mountains if they're nearby, anyplace in the outdoors that makes you feel calm, refreshed, and happy. Breathe in every sensory detail: the warmth or chill in the air, the smells and colors and textures of your surroundings. Then pick up your balloon, think about the truths of who you are, and send the balloon up to the sky. As you let it go, know

that it will join a whole community of balloons that have carried the truths of other daughters I've worked with. Watch it
ascend and feel your spirit and strength lifting inside of you.

You're much better, much smarter, much more courageous than
you were told. And you can carry all of those strengths into the
work to come.

Chapter 8

Acknowledging the Painful Feelings

ᴡ

"It feels so good to get it all out."

We've looked closely at the hidden beliefs that have pow-ered the destructive patterns in your relationship with your mother. Now it's time to examine the feelings they produce, the intense undertow of emotion that pulls you into self-defeating behavior.

This work requires great courage—a willingness to enter parts of your inner world that in some cases hold lifelong pain, disap-pointments, fears, and anger. Bravely acknowledging this material and bringing it into the bright light of consciousness will drain its power over you, and the liberation that results can be life-changing.

In this chapter, I'll open the doors to my office and let you watch as I guide other wounded daughters through this process. The emo-tions that are likely to come up can be intense, and for that reason if you decide to try the exercises in this chapter on your own, I encourage you to connect with a strong support system before you begin, trusted people who can calm, comfort, and encourage you. Even as you're reading, if you feel overpowered by emotions that come up, stop and take a break. Breathe deeply. Drink water. Go for a walk. Take things at your own pace—there's no rush.

As I've said, you may find counseling especially valuable in doing this work. A good therapist's office is a safe setting that will free you to go as deeply into your emotions as you need to, to produce lasting changes.

The Moment of Truth

Of all the tools I've used in working with people, I've found letter writing to be the most direct and effective way to get to the core of a woman's relationship with her unloving mother. In a series of letters, a daughter can tell her story fully and lay out her emotional truth without fear of criticism, contradiction, or interruption. The first letter I ask daughters to write their mothers is not to be mailed, and I ask women to write it after our first session and to read it to me the next time we meet. It's important to share the contents of the letter with a trusted person, a process that both lightens the emotional load and demonstrates the power of speaking one's truths aloud, even when they're difficult.

Many of my clients have reflected seriously on their pasts, and believe that they're somewhat in touch with what happened to them. But their letters always bring a new clarity. A letter like this is so personal that I ask women to write it by hand if they can, just to experience seeing their words in their own handwriting. Many of my clients use a computer, but I believe that holding a pen and putting the words on paper can take the writer more deeply inside, and move her truths from her hand, through her arm, to her heart.

I've designed the letter in a structured way to make it easier for daughters to get to the core of their negative experiences and what continues to haunt them today.

This letter has four parts:

1. This is what you did to me.
2. This is how I felt about it at the time.
3. This is how it affected my life.
4. This is what I want from you now.

I'll explain each part in detail, and also show you excerpts from the letters my clients read to me in our sessions, to give you a better sense of the range of memories, beliefs, and emotions that emerge in this seemingly simple exercise.

Part One: This Is What You Did to Me

My client Emily, who had struggled all her life with feelings of rejection after being raised by a cold, distant mother, approached the exercise with trepidation. But she promised to try because she was eager to get to the root of why she always seemed to wind up with men like her boyfriend Josh, who kept pulling away. (You saw Emily's earlier sessions with me in the chapter on mothers who neglect, betray, and batter.)

EMILY: "I don't know what I'll say. On the one hand I feel as though it's all a very old story that I don't need to get into again, and on the other hand I kind of like the idea of just laying it all out there once and for all."

I urged her to simply dive in. "As you sit down to write, remember: This is your moment," I told her. "It's a time when you can tell exactly what happened and get all the experiences, feelings, and thoughts that have been floating in your head for so long out into the open where you can see and work with them. You'll find that the demons of self-blame, guilt, and shame start to lose their power when you take them into the light."

The healing process kicks into gear with the words "This is what you did to me." That statement is not gentle or polite; it's absolutely direct. In fact, I know that seeing it might feel like a punch in the stomach. I deliberately removed the distancing veil of "objectivity" from the words "This is what you did" by adding *to me*. This is personal, and acknowledging that in words, and on paper, goes a long way toward freeing women to see and accept their experiences.

"Your mother's behavior hurt you," I told Emily. "Spell it out, starting with that bold, honest indictment: *This is what you did to me*. Tell your story and don't minimize it. I don't care how graphic you get. Put it all down. Did her behavior hurt you? How? How did she devalue you? What was your childhood like under her roof? Were you afraid of her? What burdens, secrets, and shames did she load you up with? You'll need to overcome your feelings of guilt and disloyalty for saying these things about your mother, but I know your desire to make your life better is stronger than your fear. Things that were extremely important and harmful may seem small to you because you've pushed them down so much, so write down the 'small things,' too. You'll gain perspective on them when you see them on paper."

Emily's eyes widened, but she nodded and said she'd give it a go.

I know that women who had violent childhoods may have a simpler time identifying their mothers' behavior as hurtful—and again, I want to caution you that if you were abused, you should definitely not confront these memories without a therapist's support. Abuse and overt bullying can seem easier to single out and describe than quieter forms of unloving behavior. But the pain and effects of unloving mothering are intense, whether it involved control, criticism, the steamroller of a mother's narcissism, emotional abandonment, or being forced to be a caretaker.

REAL EXAMPLES OF "THIS IS WHAT
YOU DID TO ME"

Emily painted a vivid picture in her letter:

"Mom—You were so critical, there was no real bond of kindness. You would never let me hold your hand or tell me you loved me. You told me once that the only reason you had me was because abortions weren't legal when you found out you were pregnant. Anything positive you did for me was done for show. You never asked me how I felt, if I was okay, what I was interested in. . . . I could never be what you wanted me to be. You used to ask me, 'What would you do if I wasn't here when you woke up.' I know you wanted to hear, 'I couldn't stand it. I'd die without you.' But I was just a scared little girl who needed her mother, and all I could think of to say was, 'Who would feed me? Who would take me to school?' And you took that as proof that I only thought of myself and wasn't worthy of your love.

"When I was older, you never encouraged anything I showed interest in, and if I didn't get the grade I wanted or the boyfriend I hoped for, you told me it was my fault and I was doing something wrong."

Emily paused as she read.

EMILY: "Am I just being a big baby, Susan? It feels so good to say all this, but I know I should just get over it."

I emphasized that it was vitally important not to minimize the pain she felt and still feels. "Don't worry that you're 'wallowing in self-pity,'" I told her. "You're not 'just feeling sorry for yourself.' It's about time you gave yourself permission to feel sorry about the things you missed out on."

It's not uncommon for women to find that the act of writing gives them access to memories they had pushed away. For my client Samantha—the daughter of a sadistically controlling mother—the first section of the letter was a revelation. Samantha had struggled with explosive anger on her job as a pharmaceutical sales rep (you saw her earlier sessions with me in the chapter on control freak mothers), and as she wrote, she flashed for the first time on how her mother had not just been controlling—she'd been abusive as well.

FROM SAMANTHA'S LETTER: "Mom, the things you did to me when I was so much younger and more vulnerable were so painful that I actually forgot a lot of them. I just remembered how you slapped me across the face when we were on vacation, for no plausible reason. I think you didn't approve of the way I was eating my spaghetti. And now I remember spitting blood after you punched me another time. I think I even lost a tooth, and the fact that it was a baby tooth just goes to show how young I must've been."

I advise daughters to stop and seek support if memories like that surface as they're writing. They're not uncommon. The act of writing is valuable in part because it can provide access to material that was so painful, it couldn't be kept in conscious memory but was shelved in the unconscious. In the course of this work, the door to that storeroom may swing open and reveal glimpses of what's long been hidden.

Some of what's buried there may be intense anger.

FROM SAMANTHA'S LETTER: "I still remember sitting in my room hating you for not letting me go to my basketball championship in junior high. Shit! There was no plausible reason for not letting me go."

Part Two: This Is How I Felt About It at the Time

Strong feelings inevitably come up as daughters look at how their younger selves were treated. So the second portion of the letter is devoted to looking closely at how they felt as girls and young women when faced with a mother whose actions made it clear that she would not or could not act in a loving way.

Feelings are the language of the heart, not the mind, and they can generally be summed up in one or two words. I felt: sad, furious, lonely, terrified, ashamed, inadequate, silly, ridiculed, unloved, terrified, angry, burdened, exhausted, trapped, bullied, manipulated, ignored, worn down, devalued. I never felt: worthwhile, smart, safe, carefree, happy, important, loved, cherished, respected.

The distinction between thoughts and feelings may seem obvious, but I emphasize it because so many people are in the habit of putting distance between themselves and their feelings by intellectualizing. That happens when "I felt" becomes "I felt *that* . . ." The word *that* carries you right into your thoughts and beliefs, and away from your feelings.

FEELINGS: "I felt unloved."
THOUGHTS: "I felt that you didn't care for me."

FEELINGS: "When you made me do all the cooking and take care of my siblings when I was only eight, I felt overwhelmed and bewildered and resentful."
THOUGHTS: "When you made me do all the cooking . . . I felt *that* you must think I was responsible enough to do it, but I felt *that* it was a lot to ask of a little girl."

REAL EXAMPLES OF "THIS IS HOW I FELT ABOUT IT AT THE TIME"

For most daughters, vividly recalling what their mothers did to them pulls them into a stream of feelings, and this part of the letter is designed to help them stay with those feelings for a while rather than pushing them away. It's not uncommon for daughters to find that they're veering into describing thoughts, not feelings, as they write, and that's okay. But the goal is to keep returning to feelings, as Emily and Samantha do below. As I reminded Emily, "If you see yourself getting stuck in 'I felt that,' go back to those statements and ask: How did that make me feel?":

FROM EMILY'S LETTER: "I felt so alone. My heart was always hurting. I felt helpless, unlovable, unwanted, unheard, and angry. I felt like I was a burden and should never have been born, and that made me feel so sad and guilty and alone. You have always been a source of pain in my life. I have always felt your resentfulness of me and my very existence. It made me feel so unloved. I hated that."

SAMANTHA WROTE: "When I was little, I just felt helpless, bewildered, confused, and really, really frightened. I felt so alone. The older I got, the more I felt angry and ashamed for what you did to me. I felt especially furious when people who didn't know your violent, sadistic side told me what a nice person you were, how funny and charming you were. I hated you for that because at home it was gloomy, depressing, and frightening. I felt like such a loser. I also felt like I had to keep a low profile and pretend to be doing fine at all times. I felt so isolated. I couldn't let anyone in."

I tell my clients to avoid self-censorship and perfectionism. This letter isn't an entry in an essay contest. What's crucial is unearthing and expressing the emotional truth. Every remembered feeling is valid and important to look at, and the intensity of some emotions can be surprising. There's no need to press on if they begin to feel overwhelming, I tell daughters. There's no rush. But honesty, as much as it's possible to muster, is vital. Recognizing, naming, and facing the emotional demons that have been in charge for so long take away their power. Sentence by sentence, the letter helps disarm them.

Part Three: This Is How It Affected My Life

This is probably the most important section of the letter. It focuses on the connections between what happened to a daughter as a little girl and the choices she's made since then. Most of the daughters we've seen in this book have reenacted much of what they grew up with without realizing it, and those are the patterns that pop to the foreground here. When I think of the links between childhood hurts and the difficulties of adulthood, I picture a long, thick rope that ties a daughter to her past and keeps her from finding the full measure of love and confidence and trust and happiness that she is entitled to. But with effort and consciousness, she can weaken that connection. Each part of this letter cuts another strand of the rope.

This part of the letter deserves thought and time. The instructions I give my clients go like this:

Describe the negative, even poisonous, lessons you learned from your mother and how they've affected your personal life, your professional life, and your life with yourself. What did your experiences with your mother teach you about your place in the world? How did they affect your sense of personal value and dignity? What did you learn about whom you can trust? What did you learn about love? Think about the self-defeating choices you've made in life and how the lessons from your mother have shaped them. You'll make vital connections between then and now.

REAL EXAMPLES OF "THIS IS HOW IT AFFECTED MY LIFE"

Many of my clients are concerned that their letters are too long, and that reading them aloud will take up most of their session. In reality, a ten-page single-spaced letter takes a little over five minutes to read. Emily, despite her initial misgivings, found that once she started writing, she didn't want to stop. She and the little girl inside her, whose pain had been so invisible to her mother, demanded their say. Emily's letter to her mother was nine pages long, and her "how it affected my life" section took up nearly half. Here's a sample of what she wrote:

"I have always lived on the fringe, like a girl looking into a playground, but never feeling as though she can participate—she is lost, disconnected, and alone. Nobody will ever stand up for her—there is no one on her side.

"I was starved for physical contact and needed to be needed. I got involved in unhealthy relationships and hated myself for doing it. I confused sex for love in a relationship, and I attracted weak men, emotional boys, perennial adolescents who refused to grow up. They had low self-esteem and little ambition. I thought I could change them. . . .

"I'm constantly thinking: 'What do others want? What do they think? What do I have to do or say to make sure they are happy?' I put my needs and wants secondary to that of others. It left me drained of energy and fatigued. . . . I feel as though I do not know how to be an adult. I have no foundation, no role models, no idea of how to set boundaries. I am terrified that people will see me as the disturbed person I must be, having been brought up by a cold, disturbed person."

Here, in this section of the letter, we demolish the argument that goes "your childhood troubles with your mother are all in the past" and the advice that urges you to "just get on with your life."

Daughters like Emily are often surprised to see how little trouble they have describing how their programming has affected them and how it fuels the compulsions they've felt to repeat many of the unhappy events of childhood in adult life. The characters and settings change, but it's as though there's just one dysfunctional tune stuck on repeat in their minds, driving a dysfunctional dance that never seems to change. Futile attempts to squeeze love from an unloving mother in childhood reappear in adulthood as they struggle desperately to prove that they're worthy of closeness, respect, and affection.

The very personal way in which each daughter falls into such patterns becomes clear as they write.

Here's a passage from Samantha's letter: "You yelled at me so much that I've always been scared to say what I wanted or demand things from people. I didn't think I had that right. I was such a sucker for the approval of others. I have become accustomed to taking things very seriously. I am constantly occupying my mind with problems and unable to live in the moment and enjoy what is going on in the here and now. My life has become just as dull and serious as yours. . . .

"I have a hard time being immune to your demands, and I always feel guilty when I go my own way or do things you don't approve of. I am so angry with myself for not having told you to leave me alone years ago. It's like there is an invisible string tying us together and keeping me from getting on with my own life."

The advice I give to clients who get stuck on this part of the letter goes like this: Keep looking for the tendrils that stretch from your mother into your life, the ones that have kept you enmeshed no matter how much you've tried to get away. This is hard work, and dumping out the stories of your life in a heap in front of you can feel daunting. Remember there's no need to *relive* the experiences you're describing. What we're doing now is looking back and *remembering*.

Part Four: This Is What I Want From You Now

The first three parts of this letter have spelled out the vivid details of a mother's unloving behavior and the lasting harm it's done. They document the outsize amount of influence she had and continues to have, and detail the power she has claimed in her daughter's life.

That balance of power shifts with the words *This is what I want from you now*. With that statement, daughters step into the role of adults who can shape their own lives. Adult daughters are not helpless and dependent anymore, and putting into words what they want from the person who hurt them so much is the beginning of empowerment.

Many daughters have not decided yet exactly what they want their mothers to do, and how—or even if—they want the relationship with them to go forward. It doesn't matter at this point. This is a first step, and there will be plenty of time to zero in on the options and come to more clarity about the decision. Nothing is set in stone, and a daughter has the right to change her mind.

The instructions I give my clients for this portion are simple: Just go from where you are now and experience what it feels like to state your preferences in an honest, direct way. I know that this may be scary, and I know that many daughters may never have given themselves permission to even *consider* changing the relationship with their mothers, because they didn't think they had the right to do it. But the time has come to shift the balance.

I remind daughters that they have the right to decide what they want, regardless of what their mothers have taught them, and despite the admonition of relatives and friends to "Respect your mother." I ask my clients to imagine that the sky's the limit and anything is possible, then answer the question: What do you most want from your mother? There's no need to have a plan or a strategy at the outset. The first step is zeroing in on a desire, knowing that it will change and evolve. What have you longed for? I ask them. What would finally make you feel free?

It may be an apology. It may be nothing. You may want your mother to stop interfering in your life, and you may want her completely out of it. The choice, I tell my clients, is *yours*.

Examples of "This Is What I Want from You Now"

Many daughters struggle initially with this part of the letter, but all of my clients have managed to sketch out a request, a desire, or a demand. Here are some examples of what they've written:

TO AN ENGULFING, OVERLY ENMESHED MOTHER: "What I want from you now is that I get to tell you what is okay to talk about and when we will see each other. In essence, I get to live my life as a normal adult. Or we will end our relationship, which will be sad and hard."

TO A COMPETITIVE-NARCISSIST MOTHER: "For years I thought that I wanted your approval or for you to change and for us to have a healthy relationship. But amazingly, today, I want nothing from you. I just want to be left alone. I like who I am today and am working with a therapist to feel solid, whole, loving, and deserving again. I have no space in my life for you. I am also rebuilding my relationship with my siblings, and know that if I let you back, you will destroy it again. I wish it could have been different between you and me—I tried to make it so. But unless you become an active participant in healing what has happened between us, then I will have accepted that it is not going to be. I have given up the illusion of a close, loving family and parent and now give my love and attention first to myself and then to the world."

TO A COLD, WITHDRAWN MOTHER: "What do I want from you now? Nothing. Nothing at all."

TO AN ALCOHOLIC MOTHER: "What I want from you now more than anything is for you to LET ME LIVE MY LIFE. Leave me alone. Go get friends, hobbies, whatever you want. Stay in your room and be depressed every day. Drink yourself into a stupor. I don't care. Just don't call me or ever try to find me. For thirty-eight years I have tried everything I can think of to live my life and still remain in contact with you, and it doesn't work. You can't stop drinking or saying hurtful things. It is beyond you. So get out of my life and let me live it however I see fit. Get out of my heart, get out of my thoughts, and just live whatever life you want."

TO A CONTROLLING, CRITICAL MOTHER: "Mom, I want you to acknowledge that you terrorized a small, defenseless child and you created a lot of damage to my soul that was really difficult to repair. I really wish you had the guts to apologize for what you did and to admit what a coward you were. I want you to see the strong, successful person I have become and understand that I have done it in spite of you, and not because of you. I want you to understand that I will do nothing more to gain your approval. I will do things my way whether you like it or not."

I caution my clients to watch out for language that gives their power away by asking for permission or approval. Notice how the woman above, writing to her alcoholic mother, says: "I want you to *let me* live my life." That sounds innocuous, and we say things like that every day, but I pointed out to her that the phrase "let me" makes her mother the warden of her life, and hands her the keys.

A far better way to express this is to say: "I'm going to live my life *my* way. . . . Without asking for your permission." That little shift makes a huge difference.

The Power of Giving Voice to the Letter

Writing the letter brings a lifetime of memories and feelings to the surface and lets daughters examine them. That in itself is healing. But writing is only 50 percent of the work. Reading the words aloud is the other 50 percent. It releases them into the air where a daughter can literally hear them—hear herself and her truths.

It is equally important that she is sharing the truth of her life and the strength of her desire to change with someone else. It's essential that those words be received by a person who can listen without judgment, without discounting, and without disbelief. A therapist is the obvious choice, but a loving partner can also serve this valuable function. It's vital that the listener and witness be chosen for his or her compassion. In the reading, and the listening, come enormous strides toward regaining what was stolen from a daughter as a child.

Chapter 9

Tapping the Wisdom
in Your Anger and Grief

ᵥ

"I'm ready to face the feelings I've pushed
down for so long."

Many intense emotions come up for daughters as they write to their mothers, laying out the facts and feelings of their history. Many significant insights surface in the process as well. Some therapists believe that insight is everything—that after the big "aha," shifts and relief will come quickly and easily. But unfortunately, it's not that simple. The truth is, dispelling the ghosts of the past requires navigating difficult emotional territory.

Most daughters find themselves grappling with a mixture of grief and anger as they face the truths about the woman who raised them. One of those two emotions is generally a familiar companion. Some women have a PhD in sadness and have long been in touch with the great sorrow that inevitably comes with having had inadequate mothering. They often tell me that the pain of feeling so little love from the woman who should've cherished and protected them was intense as they worked on their letters, and they cried many tears.

Other daughters have a powerful sense of anger, even rage, when they think of the injustice of how they were treated and how much

joy and security were stolen from them when they were young because their mothers were so singularly focused on themselves.

What I point out to the women in both camps is that although
the emotions seem quite different, anger and grief are two sides of
the same coin, and one often hides the other. Healing requires the
extraordinary power of *both* of these emotions, in equal measures.
A daughter who wants to create a life based on her own needs, not
her mother's, will have to meld the fire in her anger and the vulnerability in her grief, allowing them to create a new kind of resilience
and strength. In this chapter, you'll see how I work with my clients
to do that.

One important requirement for gaining full access to those
emotions is neutralizing the guilt and shame that unloving mothers
have nurtured in their daughters, who have long assumed responsibility for the way they were mistreated. I'll show you how we lift
the burden of misplaced blame and dismantle the beliefs that support it, as well.

If you feel strong enough to do some of this work on your own,
remember that even though anger and grief are very strong emotions, you are in control. Go at your own pace as you use the techniques and exercises that follow, and always stop if you feel shaky.
You have all the time you need to master these new skills.

Finding the Anger Behind the Grief

Allison had come to me after realizing that she'd once again fallen
for a "fixer upper" of a man who had taken advantage of her strong
tendency to rescue people. We'd traced that tendency back to the
years of training she'd gotten in caretaking while growing up in a
role reversal with her depressed mother. (You saw our earlier sessions in the chapter on mothers who need mothering.)

Her letter to her mother detailed the way she'd been expected to keep the household running from the time she was tiny, and the way her mother had leaned on her. It also described the price Allison had paid for always holding everything together and downplaying her own feelings. When she finished reading it to me, she was tearful. "I feel so exhausted, thinking about all I had to do as a kid," she told me. "That was so much to lay on a little girl."

"I know, Allison," I told her as she wiped away the tears. "There's a lot to be sad about." We sat silently for a moment, and then I asked her to think about other feelings she might be experiencing in the wake of reading her letter.

ALLISON: "Not a lot else. Not really . . . I'm just so tired, so sad. I want to scoop up the little girl I was and rescue *her* so she doesn't have to take care of anyone for a change."

SUSAN: "I think she'd be relieved. In your letter, she was pretty upset about all she had to do. Let's take another look at the section where you described it: 'This is how I felt about it at the time.' There are quite a lot of feelings in there."

ALLISON (scanning the letter): "More than I thought. . . . I was lonely . . . sad . . . I really resented my mom at times, I even hated her, and I felt so guilty about that. And when I kept having to give up activities and stay home to take care of everything, I was so, so angry too."

I'm always amazed at how clearly daughters describe their emotions at that spot in the letter, and by how many suppressed emotions surface there. These letters often provide a map to the adult daughter's inner world.

"I think many of those feelings are probably still there inside you," I told her. "They don't just disappear. You can free up a lot of energy by taking a look at those emotions again, and letting them out."

I asked her what she thought would happen if she ever allowed herself to get angry, and she answered with words I'd heard many times.

ALLISON: "I don't know. . . . I'd probably really lose it. I'd probably look hideous, and I'd lose all my dignity. I don't know if I could calm down once I started—I'd probably stay angry the rest of my life. I'd be a bitch—and nobody likes an angry woman."

Many women falsely believe that anger is a dangerous and uncontrollable force. But like a red warning light on a car dashboard, it's actually a strong signal that something is wrong—and that something needs to change. It flashes when you've been insulted or taken advantage of, when your needs are going unmet, and when someone has trampled on your rights or dignity. The healthy response is to stop in the moment and ask, "What just happened? What's wrong? What needs to change?"

Daughters like Allison, though, are accustomed to pretending their anger doesn't exist, which is the emotional equivalent of putting a piece of tape over the warning light so we don't have to feel the discomfort of seeing it. Our emotional intelligence goes unheeded, and important parts of our lives—often our boundaries and self-respect—break down. If you tend to stuff your anger, as Allison did, you're probably familiar with the results:

- Your needs continue to go unmet, and your rights and dignity continue to be ignored.
- You may turn your anger inward into physical symptoms or depression.
- You may "self-medicate" your anger with food, drugs, sex, or alcohol.
- You may resign yourself to your situation and make your anger a seething part of your identity, becoming the resentful martyr, the one who suffers, in your home or workplace.

It was time for Allison to challenge her fears of strong emotions and move the anger she was so afraid of out of her head and body so it could serve its intended purpose. I put an empty chair in front of her and asked her to imagine that her mother was sitting there.

SUSAN: "Close your eyes and picture your mother at her most helpless, demanding, and insistent. Let yourself fully conjure up the woman whose hurtful and unloving behavior you described in your letter. You're safe—it's okay. Rather than pushing the anger away, let it speak.

"To begin, start some sentences with the words 'How dare you,' finishing with whatever she did that distorted so much of your childhood. Let's give voice to that powerless child you were and the frustrated adult you've become. Let them finally have their say."

ALLISON (tentatively): "How dare you make a little girl take care of a whole family.

"How dare you think it was okay to ask a tiny child to cook, clean, take care of her siblings, and give up her childhood for you!"

Her voice began to gain force. I told her she was doing fine and encouraged her to keep going.

ALLISON: "How *dare* you suck me into your sick and twisted game with my father! How *dare* you make me your counselor! I always had to be the peacekeeper, and then you'd throw me under the bus as soon as you reconciled with him!" (She was speaking more loudly now.) "How dare you take away my happiness! How dare you infect me with this need to make you happy, even though I know it's impossible! How dare you make me feel like a failure because I couldn't fix your life! It wasn't my job! You were supposed to help *me*. How dare you turn me into a people-pleasing magnet for men who can't take care of themselves! How *dare* you!"

She paused, looking almost stunned, and I asked her how she felt.

ALLISON: "Less like a victim. I actually feel stronger."

Allison's anger was giving her clarity and conviction. I could see her gaining power with each "How dare you!" as she allowed her anger to become an integral part of her. She was using it now to express years of hurt and frustration, and at last she was facing her anger with courage instead of pushing it down and trying to avoid it.

"Hold on to that feeling," I told her. "All that anger has energy, and when you let it out, you also got a lot of certainty about what's bad for you and what you don't need to accept anymore. Notice that the world doesn't come to an end when you express your anger, even if you yell. You need to feel the heat of this emotion, and get comfortable letting it out in safe ways like this. As you do, you'll find the relief that comes from saying things you've been stifling, things you've always wanted to say. It's a kind of unburdening that ultimately will make you feel lighter."

When daughters show their anger, and listen to the valuable cues it offers, they gain access to an important component of their emotional guidance system.

Unknotting Anger to Find Grief

Samantha seemed quite different from Allison on the surface. She expressed her rage toward her sadistically controlling mother strongly as she read me her letter, almost shouting as she reached her final words and telling her mother, "I will do things my way whether you like it or not."

"I can't tell you how good it feels to finally, finally stand up to her, even if it's only in a letter," she told me when she had finished reading. "I've looked and looked at those words since I wrote them, and I think they're really going to help me change things with her once and for all."

I told her I knew that was true. In speaking her truth, she had come to genuinely see and feel that she had power, that she wasn't a four-year-old anymore. "I really noticed the humiliation and the pain of that little girl you described in your letter," I told her. "Where did those feelings go? What do you think happened to that little girl?"

SAMANTHA: "I don't know. . . . I guess she grew up and became me."

And the grown-up Samantha, I told her, still harbored layers of emotions that she'd never outgrown. All those years of humiliation and pain don't magically vanish. The wounded child is an energy that's still alive inside daughters, and that child is still afraid of being hurt. Samantha's blowups at work, and her increasingly hair-trigger temper, were a good indication of that, I told her. People often defend themselves against feelings of deep vulnerability with explosive anger.

"One thing you said in your letter sticks in my mind," I told her. "You told your mother: 'I feel like there's an invisible string tying us together and keeping me from getting on with my life.' Those old feelings are the string. They make you behave in ways that hurt you, and they come up when you least expect it."

SAMANTHA: "What do I do about that?"
SUSAN: "Let's spend a little time comforting the hurt little girl inside. She needs to know that she's safe, that no one will

ever hurt her again. Once she feels safe, you'll feel safe,
too. I want you to imagine that little girl sitting on your
lap. Imagine your arms around her. She's hurting, and she
really needs your comfort. Say what you would have wanted
someone to say to you when your mother hit you in the face
or bullied you. Start with 'Honey, I'm so sorry those bad
things happened to you. . . .'"

SAMANTHA: "Honey—I'm so sorry those bad things happened
to you. I'm so sorry Mom was so mean to you." (Samantha
stopped and looked at me.)

 "This is really hard—I don't know what to say . . . I feel so
uncomfortable."

I told Samantha that was understandable. She had been defending against these feelings, which made her feel weak and vulnerable, for a long time. Now her walls were coming down and she
felt exposed. I encouraged her to keep going, and to say what she
would say if she had adopted a little girl who had been mistreated.
I prompted her gently.

SUSAN: "You were a precious, sweet little girl and you didn't do
anything bad."

SAMANTHA: "Yeah—I like that. You were a precious sweet little
girl and you didn't do anything bad . . . I want you to know that
I will take care of you. . . . I won't let anybody hurt you or scare
you or punish you in really mean ways for no reason at all. . . .
You're safe now. You have a good mommy now. . . . You can just
be a little girl and you don't need to be afraid anymore." (She
stopped again and looked at me.)

 "Why couldn't my mother have said those things to me,
Susan? Why couldn't she have loved me like that? God, Susan,
I don't think she ever loved me at all. She couldn't have loved

me and done those things to me—that's not how someone who loves you behaves. . . ."

Her eyes welled up with tears.

"I'm afraid that's true, from all you've told me," I said. "Love doesn't make you feel terrified or lost or alone. It doesn't punish you for no reason, or berate a little girl for acting like the child she is. You're right, Samantha, what you've been describing isn't love."

SAMANTHA (through tears): "You said I was supposed to be comforting my little girl, Susan. But now I have no idea what to say. . . . I feel so completely sad and abandoned. . . . I don't know if I can go on."

SUSAN: "Samantha, I know how much this hurts. Realizing that your mother couldn't love you is one of the most painful discoveries you'll ever make. You deserved to be cherished, but your mother was a disturbed, unhappy woman who took out her frustrations on you. And it wasn't your fault. It wasn't little Samantha's fault. She was innocent. Nothing you could have done would have made your mother love you more. She couldn't. We can't know why. But we can be sure of one thing: It had nothing to do with you. It wasn't your fault. I want you to say this back to me, Samantha: 'It wasn't my fault.'"

SAMANTHA (very quietly): "It wasn't my fault."

I took her hand and held it as she wept.

"Say it again, louder," I told her.

SAMANTHA (stronger): "It wasn't my fault."

SUSAN: "Louder. Make me believe you."

SAMANTHA (yelling): "It wasn't my fault!" (She took a deep breath and looked at me.) "Susan, it *wasn't* my fault."

It Was Never Your Fault

"It was all my fault" is the pervasive lie daughters *must* refute if they want to reclaim their lives. It's a belief that leaves them laden with guilt—and the sense that they're responsible for their own mistreatment and their mother's lack of love. Deeper than that, it's a powerful source of shame—the feeling that says, "There must be something wrong with me, or this wouldn't have happened."

Implanted by a lifetime of programming that may have been punctuated by slaps, sighs, criticism, finger-pointing, or yelling, the belief that daughters are responsible for their mothers' choices, feelings, and behavior toward them is the force that holds dysfunctional mother-daughter relationships together in their destructive status quo. As long as daughters feel guilty and shameful for having "caused" their mothers' unloving actions, they're incapable of challenging them, standing up for their rights, or fully experiencing their anger and grief at what they've experienced. They continue to accept the lie that they are bad, selfish, wrong, and defective, and therefore have no right to feel what they feel, or do what's right for themselves, regardless of what their mothers want. "It was all my fault" is the big lie that enables all the other lies unloving mothers tell—the ones my clients challenge in the Lies and Truth Exercise—to flourish.

Challenging this last big lie until it loses its hold brings daughters a freedom, and a sense of self-love and self-acceptance, that's often been missing all their lives.

There's a great deal of power in the words Samantha spoke: "My mother couldn't love me *and it wasn't my fault.*" Repeating that several times, until the words sound and feel convincing, has been a giant step toward liberation for many of my clients. Simple though the action is, speaking that truth aloud seems to carry it deep inside.

I suggested to Samantha that she take the work one step further by telling the little girl on her lap what *she* wasn't responsible for. "It can be very healing and comforting to do this," I told her, "and it will go a long way toward freeing Little Samantha from the huge load of guilt and shame she's been buried under all her life."

SAMANTHA: "Okay. . . . Sweetie, I know Mom always told you you were bad, and you deserved to be hit and punished and yelled at. She told you it was all your fault. But she was wrong. You were a wonderful, smart little girl. And you weren't responsible for her slapping and punching you. You did nothing to deserve having a tooth knocked out. You weren't responsible for the way she kept you from going to the basketball tournament. You didn't do anything to deserve that. You weren't responsible for her cruelty. You weren't responsible for the way she was obsessed with your studying and grades. You were a smart little girl. You didn't deserve to be treated as though you were lazy and stupid. You weren't responsible for Mom's craziness. It was all hers. It didn't have anything to do with you."

When a daughter pictures the child she was, and remembers the burdens and unloving behavior that this little girl had to face, it's far easier to accept the truth: The child was blameless. The woman she's become is blameless as well. The reasons given for abuse, mistreatment, role reversals, and suffocating behavior had everything to do with her mother, and nothing to do with her.

When that truth penetrates, as it continued to do over time with Samantha, a great deal of emotion can surface. I asked Samantha how she was feeling.

SAMANTHA: "So ripped off and outraged . . . but mostly incredibly, horribly sad. No—it's beyond sad. I feel like somebody died."

Samantha was weeping now, grieving. And her grief helped carry her to a deep acceptance of the truths that were clear to her now. She could no longer call her mother's unloving treatment love. She wasn't responsible for the terrible treatment she'd received—her mother was. The world seemed turned upside down.

Everyone who comes to this point must grieve the enormous losses they have had. The majority of daughters of mothers who couldn't love them lost a great deal of their childhood because their mother didn't allow them to have one.

I enumerated those losses to Samantha:

"You lost the right to be playful and silly.

"You never got to feel the joy of being a carefree kid. You were four, or fourteen, going on forty, bearing adult burdens from the time you were small.

"You didn't experience the predictability, consistency, and nurturing that could give you an inner sense of security.

"You had far too few opportunities to feel free and trusting.

"You were starved for validation of your worth, and your mother hid the truth from you: That you are a unique and wonderful person whose job in life is to be herself.

"You've kept this knowledge hidden inside on a semiconscious level for most of your life," I told her, "and at last it's in front of you, where you can take it in fully with your mind and heart."

Guarded and wary, Samantha had cut herself off not only from other people, but also from her own soft and loving feelings. I promised her that they, too, were on the other side of the pain she was experiencing now.

Putting to Rest the Fantasy of the Good Mother

Samantha was, naturally, very raw at this point, and I suggested that she might want to do an exercise that would help calm her and put her in a better emotional place. The exercise would be a symbolic burial of the fantasy that had kept her, and so many unmothered daughters, running after the love from their mothers that would always be held just out of reach. "Are you up to trying it?" I asked.

SAMANTHA (with a slight smile): "Yeah, it can't get any worse, that's for sure. I've come this far—let's go for it."

I picked up a small bunch of dried flowers that I keep in my office for this purpose and put it on the coffee table. "Imagine that this table is a coffin," I said. She flinched at the word *coffin*, and I assured her that this was all symbolic. "Now visualize the coffin being lowered into the ground. What we're burying is the fantasy of the good mother, so over the casket, I'll show you how the eulogy goes, and you can add your own words."

SAMANTHA: "Okay."

To get her started, I began: "I hereby lay to rest my fantasy of the good mother. It wasn't in the cards for me. It didn't happen, and I know it never will. It wasn't my fault."

SAMANTHA: "I did everything I could think of to get her to love me, but nothing worked. . . ."

"I'm no longer willing to keep banging my head against a stone wall to try to earn her love," I continued.

SAMANTHA: "I'm no longer willing to distort the rest of my life to try to make her happy. I'm not going to pretend that the few crumbs I got was real love. It's hard to let go of you, but rest in peace, my fantasy. I need to get on with my life."

Samantha sat with her eyes closed, wiping away tears with the back of her hand. I asked her how she felt.

SAMANTHA: "Sad. Still sad. But calmer. A little freer. Stronger. A little more me."

The most important thing about the eulogy isn't the words. It's the way it helps women put to rest the fervent wish that their mothers will suddenly transform.

The eulogy is a potent vehicle for change because with its symbols and symbolic behavior, it speaks directly to the unconscious and begins to reprogram it. A symbolic burial is a powerful way to end the "if onlys." "If only I do this she'll be nice." "If only I rescue her, she'll be happy." "If only I let her steal enough attention from me, she'll eventually give some back." "If only I become perfect enough, she'll stop criticizing me." "If only I try hard enough, she'll finally love me."

For daughters who have been focused all their lives on figuring out what they were doing wrong, and searching for a way to win their mothers' love and approval, the acceptance that they have been chasing a fantasy that wreaked havoc with their lives marks a turning point for them. Samantha felt exactly that.

There's No Magic Wand: Living With– and Through–Anger and Grief

My clients often swing from grief to anger and back again as they adjust to the truth. I tell them that it's appropriate and healthy to feel sad—or furious—as your whole being acknowledges that you didn't receive the love you deserved and needed from your mother. Viewed through this new understanding, a memory from childhood may unleash a flood of sadness. Irritations that may have been easy to take for granted in the past may now trigger unusual anger. It's important for daughters to take each emotion as it comes, and remember that what will bring the greatest peace and power in the long run is working with these difficult feelings rather than pushing them away.

In the period after writing the letter and experiencing the emotions that it inevitably raises, I urge daughters to focus on learning new ways to work with their feelings. This is not the time for confrontations, drama, or battling with Mother.

"It's better simply to be with the anger, the grief, and to let these feelings teach you more about yourself and what you truly need and desire," I told Samantha. "What you discover now will become the foundation of decisions you'll make about how you want to proceed in your relationship with your mother, so don't try to race past your feelings on the way to a resolution."

A Toolbox for Handling Anger and Grief

As daughters work toward finding a new path with their mothers, I remind them that they don't have to let their emotions build without release—there are many ways to express them constructively. Below, I've collected some of the tools and insights that have helped bring many of my clients relief when their feelings intensify. Wherever you are in facing your relationship with your mother, I think you'll find this toolbox helpful anytime you're feeling agitated, confused, or overwhelmed.

Confronting and Managing Anger

Even when daughters think they're old hands at expressing anger, it's likely that what they know best is the "stuff and erupt" pattern, holding in their feelings until they can't be contained. Some women go from stuffing their anger at their mothers to exploding at anything or anyone who activates old wounds or old memories. They jeopardize their personal and professional relationships, and ironically the person they're most deeply angry with—their mother—can't even hear them. Others channel unexpressed anger into physical symptoms.

Women who are actively in touch with their anger, and express it by yelling at their mothers, may think they have this tricky emotion well handled. But yelling is as useless as not saying anything. As I tell my clients, it reduces you to a child and strips you of credibility. Worse, there's no possibility for change, because she doesn't hear what you're saying once you start screaming. All you've done is to once again hand your power to her.

There are much better alternatives, which bring much more positive results. Here are the instructions I give my clients to teach them some of the most effective techniques I know for managing anger.

1. FEEL YOUR ANGER WITHOUT JUDGING IT.

I know that for some women, acknowledging anger is difficult because it makes them feel intensely guilty and disloyal. But to be human is to feel anger—all of us do. It's not a flaw, it's an essential part of our emotional guidance system. To allow yourself to feel, and be served by, this emotion, try approaching it with curiosity. Ask yourself: What is my anger telling me to look at? What needs to change?

2. ACKNOWLEDGE THAT YOU HAVE THE RIGHT TO FEEL ANGRY.

Tell yourself:

- I have been deeply hurt, and I have the right to be angry about that.
- Anger doesn't make me a bad person.
- It's normal to feel guilty about being angry.
- My anger will give me power when I manage it in a healthy way.

3. GET A REALISTIC IDEA OF WHAT ANGER LOOKS LIKE.

If one of the reasons you run from your anger is that you're afraid it will make you look ugly, pay attention to TV shows and movies in which women express their anger and assert themselves in a strong, controlled way. You'll see that their faces often take on a kind of strength and firmness that's actually attractive in many ways. They don't look like shrews; they're almost regal. One of my favorite movies is a classic film called *The Heiress*, in which Olivia de Havilland plays a plain and painfully shy young woman who's been beaten down by her tyrannical father and betrayed by her fortune-hunter fiancé. As she evolves during the movie, acknowledging her anger and the reality of how she's been treated, she actually transforms physically. In the last scene, her demeanor and the expression on her face make it clear that she will never be taken advantage of again.

The strength she radiates is the opposite of ugly. She has the beauty that comes with empowerment.

4. RELEASE THE ENERGY OF YOUR ANGER WITH PHYSICAL ACTIVITY.

Run, walk, hit a tennis ball, swim, lose yourself in the loud music and sweat of an exercise class. Moving your body will release endorphins, the vital brain chemicals that are so important to your sense of well-being. It's one of the best ways to dissipate anger that's building up inside.

5. PUT YOURSELF IN A MORE PEACEFUL PLACE WITH VISUALIZATION.

Pick a time when you won't be interrupted for five or ten minutes and sit down in a comfortable, private place—your favorite chair, the top of your bed, even your car. Close your eyes. Breathe in deeply through your nose and let the breath out slowly through your mouth. Visualize your breath as a warm, smooth current that flows in and out, and take four or five long, slow breaths. Feel the breath going into any tight places in your body, and when you exhale, let your breath carry the tightness away.

Now visualize the most beautiful, serene place you've ever been. (For me, that place is a sparkling blue bay on the Big Island of Hawaii surrounded by black-green mountains.) See yourself in your special place. Let yourself be nourished by the air, the sun, the wind, the colors, the smells. You'll notice yourself feeling calmer. Stay as long as you like, breathing it all in and soaking up the peace of this place. All you have to do here is rest and breathe. Let your thoughts float away on the breeze. Feel your heartbeat and breathing slowing down. Linger here in the quiet. When you're ready to leave, take a last look around and then open your eyes. This place is always there for you. You can return anytime you want to.

WHEN THE PERSON YOU'RE ANGRY AT IS YOU

Once a woman gets in touch with her anger at her mother, she may find that she's full of questions about how someone who was supposed to love her could have behaved with so much thoughtlessness or even cruelty. The next question that immediately follows is often for herself: How could I have continued to tolerate such treatment for so long (even into the present)?

Samantha, like most daughters, had grown up believing that if her mother was unhappy or unkind, it was because she wasn't doing things right. But now, she told me at one of our sessions, she found herself thinking: "How could I have let her mistreat me? How did I let her control me like that? Why can't I stand up to her? How could I have let that happen to me? How could I have been such a slave to trying to make her happy?" In essence, she'd walked away from her old form of self-blame only to replace it with a new one. She was still asking, "What's wrong with *me*?"

It's crucial that daughters look critically at the blame and the anger they may be directing at themselves. This is what I told Samantha, and what I tell all my clients:

First of all, you were dependent, helpless, and programmed to obey authority. Your mother was bigger, older, smarter, and much more powerful than you were, and you had to go along. You didn't have a choice. What were you going to do, leave home and get a job when you were seven?

It's natural to be upset at yourself—that's part of the process when you let yourself see just how badly you were treated. But self-recrimination is an exercise in futility. It doesn't resolve anything, and it doesn't make you feel better or improve your life. It only amplifies your discomfort.

I ask my clients to use these phrases to connect with the healing self-forgiveness and self-compassion that they deserve. Read them, repeat them, or write them down, I advise, whenever the thought "How could I have let this happen?" begins to surface.

- I did the best I could with the information I had at the time.
- As a child, I didn't have the ability or perspective to know what was really going on.
- I was programmed early on to defer to my mother and try

to please her, and that programming has been tenacious and powerful.
- My guilt and my fear of the consequences were stronger than my motivation to change the relationship, but that is changing.
- I'm not alone; many people have trouble disentangling themselves from unloving mothers.
- Change is difficult for everyone.
- It was hard for me to give up the hope that things could get better; I couldn't accept that she was very unlikely to change.
- I was so disempowered that I didn't have the tools to change the status quo.
- I forgive myself.

Not only did they not have the tools they needed to set firm limits and boundaries between themselves and their mothers, but they also didn't know they had the right to acquire and use them.

The Truth About Grief

For a time after they write their letters and face the pain of their pasts, almost anything might trigger daughters' grief. A memory. A movie on TV showing a mother and daughter sharing the kind of intimacy a woman longed for but never had. The sadness is normal. As I tell my clients, it lets you know that you're a sensitive person with feelings that you need to honor and protect.

I know how frightening this kind of grief can feel. It can be piercing for any woman to truly acknowledge that her mother was unloving. In their sadness, some women tell me it seems at times as though they're at the bottom of a deep, black river and they'll never come up for air. Some of them feel panicky as they experience the intensity. But I reassure them that they're not going crazy, they're grieving. They're not going to fall into a million pieces. Their tears are allowing them to heal.

Grief, like depression, is something we always believe will never go away. We fear that we'll feel this way for the rest of our lives, and for that reason we may pin on a smile and pretend everything's okay so we can just get on with things. Or we discount what we feel by saying, "I know people who've had it worse." We don't want to wallow in our sadness.

But if we don't confront our grief by facing it bravely, it is likely to continue to have a powerful hold over us. We have to go *through* grief, not behind or around or over but *through* it. It takes great courage to do this, I know.

I wish I could spare daughters all of this, or hand them an instant, grief-erasing exercise. But there's no such thing. I can promise them, though, that if they let themselves acknowledge and feel their deep sorrow, it will diminish gradually. And over time it will begin to lessen significantly.

The visualization I described earlier in this chapter for easing anger, and the suggestion about using exercise and movement to release emotion, can help when grief feels overwhelming as well.

Using Your Emotions to Break a Cycle

Every daughter has a pivotal choice to make as the pain of her relationship with her mother continues to mount. She can struggle through the process of coming to terms with her feelings and use them to guide her to clarity and real change. Or she can sit on those feelings and defend herself against the pain by acting in hurtful and inappropriate ways—just as her mother did.

Having the courage to stay with difficult feelings and learn all they have to teach us is a daughter's greatest insurance policy against turning into her mother. And it moves her along the path that will allow her to fill her life with the kind of genuine love her mother so rarely shared with her.

Chapter 10

Change Your Behavior, Change Your Life

🌵

"I see that change is really hard, but not changing is harder."

daughter who does the demanding work of confronting her emotions comes to know on a deep, almost cellular level that she was not responsible for the pain of her childhood. And as that knowledge replaces the guilt, shame, and self-blame that have controlled her, it becomes increasingly difficult for her to continue accepting her mother's unloving behavior, and her own self-defeating responses to it.

At this point, though, most of my clients have no idea what to do or how to carry the profound changes in their inner worlds into their everyday lives. Daughters of unloving mothers had terrible role models who never taught them their responsibilities to themselves or their rights as independent women. They got few or no lessons in how to stand up for themselves, handle conflict and stress in a healthy way, or set protective limits on other people's behavior.

But a daughter needs all those skills to break the patterns of a lifetime, and she needs new tools for becoming the woman she was meant to be.

In this chapter I'd like to show you the blueprint for a new way of being, and teach you a set of behavioral strategies and communication skills that have given my clients the resources to begin shifting their relationships with their mothers, even as they were processing the intense inner work you've seen in the previous two chapters.

Every one of us needs to master the art of using self-protective, assertive behavior. It's the most effective defense against mistreatment, and I feel comfortable telling you that you can practice it on your own, regardless of whether you're in therapy or how much meaningful emotional work you've done on yourself. This chapter is behavioral, strategic, and technique oriented. We think of behavior as the end of the growth process that involves changing feelings and beliefs, but I think you'll be surprised to see how these new behaviors will dramatically shift your feeling state and any residual negative beliefs. The tools you'll find here can help you make major changes in your life.

An Adult Daughter's Responsibilities and Rights

Once you disconnect from the belief that you are responsible for your mother's happiness and well-being, a void often seems to open up—the emptiness of the unknown. You've probably been shaping your life in response to your mother's influence since you were little, and even if you have minimal contact with your mother now, the habit of putting her desires first (or reacting against them) may still crowd out your normal instincts for self-care and direction. Now, as you think about setting your own agenda, you may be slightly overwhelmed about how to begin.

You can start with a set of new beliefs that have the power to supersede "I am responsible for my mother"—and act as a compass that will always lead you back to choices that are both self-nurturing and respectful of others. These are your *real* responsibilities, responsibilities I believe you're more than ready to accept.

As an adult daughter you are responsible for:

- Claiming your own self-worth.
- Having the life you want.
- Acknowledging and changing your own behavior when it is critical or hurtful.
- Finding your own adult power.
- Changing the behavior that's a replica of your mother's unloving programming.

You are in charge of these behaviors and accountable for them. At first, you may not know fully what it means to honor these responsibilities, or how you'll do it. That's okay. This is behavior you're aiming toward, a new destination on the map. You're now leaving the world your mother dominated, the one in which "having the life I want" and "having my own thoughts, feelings, and behaviors" were often treated like punishable offenses. You'll need to make some inner and outer shifts to adjust to the significant change you've set in motion. Reflecting on these responsibilities, letting them seep in, is an important first step, so take time to do that.

As you accept your real responsibilities to yourself, you're also ready to acknowledge your rights, a set of basic entitlements that are yours to claim as a strong woman and daughter. I created the list below on a recent Independence Day. In thinking about the holiday and the inspiring example of breaking away from tyranny that it celebrates, it occurred to me that many women, faced with the coercive and even tyrannical behavior of unloving mothers, have gone through life not knowing they had the right to protect themselves and seize their own freedom. For them—for *you*—I drafted this Bill of Rights. If you internalize and observe it, you will have withdrawn the permission for *anyone* to treat you badly.

The Adult Daughter's Bill of Rights:

1. You have the right to be treated with respect.
2. You have the right to not take responsibility for anyone else's problems or bad behavior.
3. You have the right to get angry.
4. You have the right to say no.
5. You have the right to make mistakes.
6. You have the right to have your own feelings, opinions, and convictions.
7. You have the right to change your mind or to decide on a different course of action.
8. You have the right to negotiate for change.
9. You have the right to ask for emotional support or help.
10. You have the right to protest unfair treatment or criticism.

As an adult, you have always had these rights, but after years of programming, you may not have allowed yourself to act on them. As a child, you may well have been guided in exactly the opposite direction, punished for being less than perfect. And today, you may love the *idea* of these rights but be uncomfortable, even anxious, as you imagine insisting on them. Your mother has probably been in the driver's seat for so long that even as an adult you may often feel like a little girl whose feet don't reach the pedals and who can't see over the dashboard.

You're stronger, more courageous, and much more powerful than you think. And you'll demonstrate that to yourself again and again as you learn and practice the first of several essential life skills: nondefensive communication. This skill can help you make major changes in the way you communicate and deal with conflict. And, perhaps for the first time, you can begin exercising your rights and following through on your responsibilities to yourself.

Using Nondefensive Communication

It's likely that your mother is still pressuring you to let her have her way by cajoling, criticizing, threatening, crying, sighing, trying to make you feel guilty/inferior, or bulldozing past disagreements with "Don't talk back to your mother!" or threats of repercussions. Your exchanges probably fall into a predictable pattern. She goes on the offensive, sometimes very quietly, and you are forced to play defense.

The "I attack / you defend" pattern works extremely well for your mother, because it cements her one-up position with you. You've almost certainly become an expert at explaining yourself, denying that you've done anything wrong, rationalizing, giving excuses, giving reasons, and apologizing. But you probably don't realize that every time you reach for those familiar responses, even though you think you're defending yourself, you've actually been forced onto the defensive—and there's a huge difference between the two. To defend is to protect from harm. But defensiveness signals weakness, and an eagerness to avoid challenge or criticism. It never positions you as an equal.

Here are some well-used phrases from the defensiveness play-book:

- I am *not.*
- No I didn't.
- How can you say that about me?
- Why do you always . . . ?
- Why can't you be reasonable for a change?
- That's crazy.
- I never said/did that.
- I only did it because . . .
- I didn't mean to.

- I was just trying to . . .
- But you promised . . .

Anxiety, worry, fear, and generous amounts of vulnerability are embedded in every defensive word you say.

Defensive language is your enemy. Each time you're defensive, you create an opening for your mother, and signal your willingness to be drawn into an endless loop of accusation and defense that resolves nothing. Automatically you've backed yourself into a corner and invited her to bring on the pressure. On the defensive, you look and feel weak—and you are.

But you can break this cycle, and you can do it so easily it may feel like magic—by changing the words you use.

Sharon: Standing Up to Abusive Criticism

Sharon, an MBA who was working as a doctor's receptionist, was seeing me for panic attacks that had flared up after some recent conflicts with her highly critical, narcissistic mother. (You can see our earlier sessions in the chapter on narcissistic mothers.) Writing her letter had helped her begin to see herself in a more positive light, but she hadn't been able to calm her volatile exchanges with her mother.

SHARON: "It happened again, Susan. It was Aunt Mona's birthday, so I went to lunch with her and my mother. I love Mona, and we were catching up, since we hadn't seen each other in a while. She asked what I was doing now, and before I could say a word, my mother jumps in. 'It's a tragedy,' she says. 'She's working in a doctor's office. As a *receptionist*.' She said it like I was a garbage collector or something. Then she said, 'All that education down the drain.' And with a kind of tragic little laugh she says, 'She's my little failure.'"

I asked Sharon how she'd responded.

SHARON: "You would've been so proud of me, Susan. I stood up
for myself. 'I am *not* a failure,' I told her. 'I'm proud of myself!
I like my job, the people are great, and it makes me happy.
Why can't you be happy about that? I didn't *want* a high-stress
job. Why do you always need to humiliate me?' That shut her
up for a moment, but of course she had to have the last word.
'It's all right, dear,' she said. 'I know you're sensitive about
that. But you need to straighten up. You're not always going to
have me around as a safety net.' As if she ever had been. I was
fuming. Fortunately, Mona changed the subject and Mom let
things drop."

I asked her how she felt about the exchange.

SHARON: "Not so great, to be honest. . . . I did stand up for myself,
and I thought I would feel great afterward. I know I did in the
moment. But when it was all over, I still felt like hell. And I
can't quite figure out what went wrong."

I explained to Sharon that her mother's dismissive and insult-
ing comments ("She's my little failure") had pounced on a painful
old theme—*You're not good enough*. And almost immediately, the
familiar loop of insults began to play in Sharon's head again. Her
first impulse was to protect herself and defend against further hurt.

"The problem is," I told her, "the strategies you used to protect
yourself made things worse, not better. It seems like the natural
thing to do, but every time you try to justify what you've done, or
ask 'why' questions, like 'Why do you need to humiliate me?' you're
actually giving your mother ammunition. You're almost certain to
wind up feeling small, humiliated, and less-than, even if you yelled
back."

I explained that as long as she was defensive, her mother was in control of the conversation and the agenda. Sharon's defensive responses were *inviting* more critical comments, more jabs. And as she felt more attacked and frustrated, it was easy to revert to behavior that was pretty ineffective. "I know you don't have kids of your own, but I'm sure you've seen kids fighting," I told her. "One says 'You cheated!' The other says 'I did not!' And the fight goes back and forth like a Ping-Pong game: 'Did not!' 'Did too!' 'Did not!' 'Did too!' When you're essentially doing the same thing with your mother, it's easy to come away feeling like a five-year-old."

I suggested we do some role-playing, a treatment technique I have used for many years. Role-playing is an extremely effective way to model new behavior and cut to the heart of an issue.

SUSAN: "You be your mother and I'll be you, and I'll show you some far better responses than the ones you've been using. It's easy to learn them, and they stop the critical person in her tracks, at least temporarily, and give you time to regroup. To start, pick one of your mother's pet criticisms of you. I want you to sound as much like her as you can."

SHARON (as mother): "I just don't understand how you can throw away your MBA so you can answer phones and file things. But you've never listened to me. If you had, you wouldn't be such a disappointment."

SUSAN (as Sharon): "I'm sure you see it that way."

SHARON (after a long pause): "Really? That's it? Susan, I don't know what to say to that."

SUSAN: "Exactly. Neither will your mother. When you take the defensiveness away, there's not a lot for her to hold on to. Let's try it again."

SHARON (as mother): "I hate to say it, but you've really let us down. I guess you'll always be my little failure."

SUSAN (as Sharon): "I don't accept your definition of me."

SHARON: "That's all I say? It's so . . . abrupt. Shouldn't I say
more?"

SUSAN: "No, that's all. A simple statement will do. Don't add,
don't embellish, and don't think you need to carry on the
conversation. It will probably feel awkward to do this at first,
but it will get more comfortable the more you practice."

I gave Sharon the following list of nondefensive responses and
asked her to memorize them and to practice using them with a
friend. As you try this, think about your mother's most common
criticisms and the words she uses to pressure you, then select the
responses that fit the situation. Practice them on your own until
they feel comfortable and automatic. It will take thought and effort
to use them at first with your mother because they replace the au-
topilot responses that most people rely on. But I promise you, the
results will be worth it.

Nondefensive phrases:

- Really?
- I see.
- I understand.
- That's interesting.
- That's your choice.
- I'm sure you see it that way.
- You're entitled to your opinion.
- I'm sorry you're upset.
- Let's talk about this when you're calmer.
- Yelling and threatening aren't going to solve anything.
- This subject is off-limits.
- I don't choose to have this conversation.
- Guilt peddling and playing the pity card are not going to work
  anymore.

- I know you're upset.
- This is nonnegotiable.

Ninety-nine times out of a hundred, these phrases will act like a referee coming in to stop a fight. They nip conflict in the bud. You won't need them when someone is pleasant, but they're essential when you're being blamed, bullied, attacked, or criticized.

I told Sharon that I was willing to bet that once she began using this new language, her panic attacks would significantly lessen. She wouldn't be so emotionally naked and vulnerable anymore. Now she'd have weapons. And as she pointed out when she reported back at our next session, she had a shield.

SHARON: "Mom was just flummoxed when I didn't take her bait and start defending myself. I felt pretty silly practicing my lines beforehand, but it was great to have them in mind when I needed them. With this script, I feel like I have a moat around me and she can't get to me the way she used to. I think it's really going to help. It already has."

YOUR FEELINGS WILL CATCH UP WITH YOUR NEW BEHAVIOR

You may be nervous about what your mother will do when you begin using nondefensive communication. But don't let anxiety stand in the way of putting this new behavior into practice. It doesn't matter if your stomach is in a knot or your neck is tight the first time you try it. It doesn't matter what's going on inside you. *Change your behavior and the feelings will catch up.*

The pain, humiliation, and frustration you felt when your mother had the upper hand will dramatically ease, and you'll feel your pride and power expanding. But you've got to take the first step. The new learning can't just stay in your head—you have to take action. Until now, your mother has had the power. You can change all that. Make a commitment to yourself that you won't let fear or anxiety control you, and if you slip, it's okay. You'll get it right the next time, and little by little, nondefensiveness will become more automatic.

Lauren: Pulling Off the Tentacles of Enmeshment, One by One

Lauren, a stockbroker with an overly enmeshed mother who insisted on daily check-ins and an all-access pass to her daughter's schedule, was certain that her mother would "have a cow" if she tried to regain some power in the relationship. And she was nervous about trying. (You can see our earlier sessions in the chapter on overly enmeshed mothers.)

LAUREN: "What I really want to do is stick to my guns and just *not* call her when I don't feel like it. But I know exactly what will happen. She'll call and read me the riot act, and pretty soon she'll be talking about how she was sitting in the dark waiting for me to call. . . . And I'll cave."

As I did with Sharon, I suggested to Lauren that we do some role-playing. I'd play her mother and let her play herself so we could see her usual responses.

SUSAN (as mother): "I know you asked me not to bother you at work, but I've been so worried about you since you didn't call last night. I was sitting here, wondering if you were okay, wondering if something terrible had happened to you. I tried to watch TV last night, but all I could do was think of you in a car wreck. How could you make me worry like that? What's happened to you? Have I done something to upset you? You've hurt me and worried me. I didn't get a bit of sleep last night. Don't you care about me?"

LAUREN: "God, Susan, you've been tapping my phone. . . . Okay, here goes. . . . Mom, of course I care about you. I've been calling every night for years. I just missed one night. I don't think that's a federal offense. Look at all the places I take you and all the time I spend with you. I've been a very devoted daughter."

SUSAN (as mother): "That's not what it feels like to me. Not after last night. You don't know how much I worry. . . ."

LAUREN: "I feel like a total wuss, Susan. I know we're just pretending, but I feel guilty as hell. How am I possibly going to tell her I'm not going to call after all that?"

I told Lauren that many times, healthy new behavior has to precede emotional change. When you start to say "No, I'm not going to do what you ask just because you insist," you will probably feel wobbly and guilty. But the more you do it, the easier it becomes, and the more the anxiety will dissipate. We all have to behave in healthy ways and trust that the healthy feelings will catch up. They always do.

Lauren's face lit up as we practiced nondefensive responses to her mother's pressure. She especially liked "I'm sorry you're upset" as we worked with it because it sounded kind, but it made her feel powerful.

SUSAN (as mother): "You had me so worried when you didn't call last night. I didn't sleep a minute. How could you be so thoughtless?"

LAUREN: "I'm sorry you're upset. I certainly didn't intend to worry you."

SUSAN (as mother): "You know how much it eases my mind to

have our little phone check-ins. Is that too much to ask?" (I
raised my voice and acted as if I were crying.) "Don't you care
about me anymore?"

LAUREN: "Of course I do. What a ridiculous question. . . .
Whoops. I fell into the trap. Can we try it again?"

SUSAN: "Sure."

SUSAN (as mother): "Don't you care about me anymore?"

LAUREN: "Oh God. I can't think of what to say."

SUSAN: "How about something like, 'Let's talk about this when
you're calmer.'"

SUSAN (as mother): "I can't ever talk to you. I'm going upstairs to
rest."

LAUREN: "Can I really do that, Susan? Don't I sound like a snotty
bitch?"

At first, when you begin to use your nondefensive phrases, you
may think you're doing something outrageous or wrong. You're not
used to letting leading questions and statements dangle in front of
you without reacting. And you're certainly not used to letting your
mother's upset go by marked only by a neutral phrase.

There's no way of predicting how your mother will react to
your new behavior; she's not used to being challenged effectively.
A narcissistic mother might well respond to your nondefensiveness with anger if you tap her narcissistic rage. If she's enmeshed,
like Lauren's mother, she may pull out the pity weapon. I can't
cover every conceivable reaction your mother will give you. All
I can say is this: Stay nondefensive. The list of nondefensive responses above will serve you well no matter what she does. If you
need to cut short a conversation because she has become enraged
or verbally abusive, just say, "Let's talk about this when you're
calmer. I need to go now."

This kind of communication will seem like a new pair of shoes
that pinches and feels as if it doesn't belong to you. But comfort will

come with practice. The anticipation is always worse than the do-ing, and you'll find that the relief and pride in holding your ground effectively is much stronger than your fears.

Don't worry if you slip into old defensive patterns. It will happen—no one gets this right all the time—but you'll have many occasions to try again. Keep weaving nondefensiveness into your relationship with your mother. The more you do this, the more comfortable you'll get.

It's impossible to overstate the importance of learning these skills and making them part of your everyday behavior. If you can do it with your mother, you can do it with anyone.

Chapter 11

Setting Boundaries

ᵥᵧ

"I would never have believed I had the right to say no."

Nondefensive communication effectively deflects and de-escalates conflict. It allows you to respond to even the most inflammatory statements in a cool, centered way.

To shift the balance of power in the relationship the most effectively, though, you will need to do more than just respond nondefensively. You'll have to begin defining your own needs and wants and communicating them to your mother. That requires setting boundaries—limits and rules for your interactions with her.

Imagine living in a house that had no doors, windows without panes, and a yard with no fence around it. With no secure barriers to protect your space, privacy, and safety, you'd feel exposed and vulnerable. That's what you've probably experienced all your life in your relationship with your mother. Unloving mothers don't recognize the boundaries between themselves and you—it's a trait they all share. They assume that their likes and needs are more important than yours. Some invade your physical space. All impose their judgment, priorities, opinions, and preferences on you. They take over, and they insist that that's what good daughters let their mothers do.

Setting boundaries changes all that. It allows you to define both your physical and emotional space and to declare dominion over your own life. Physical boundaries have to do with what you'll allow people to do in your physical presence and inside your house. Emotional boundaries define how people are allowed to treat you. The letter-writing work described earlier goes a long way toward helping women understand the emotional separation between themselves and their mothers, but even if you've done that work, it may still be difficult for you to separate yourself from your mother's actions and responses today. The questions below will help you assess how much you're still playing by your mother's emotional rules:

- Do you still take responsibility for your mother's feelings and needs?
- Do you still prioritize her feelings and needs while neglecting your own?
- Do you still become upset because your mother is upset?

If you answer yes to one or more of these questions, your emotional boundaries are weak. You're living in emotional territory ruled by your mother, not by you.

It is not surprising that women who have been taught that their feelings and needs can't possibly matter compared with their mothers' have weak emotional boundaries. If you grew up this way, drawing a line between your emotions and your mother's may feel wrong—and completely foreign. But it will equip your emotional world with the doors and fences that need to be in place before you can have the kind of independent life you want.

You are not responsible for your mother's life, moods, feelings, or distorted perceptions of you. They belong to her. And no matter how guilty it makes you feel, your job is to create a boundary between her life and yours.

It's essential to do this because as long as your focus remains on your mother, you can't possibly know who you are or what *you* really want. Anticipating her needs, reactions, and upsets takes the place of exploring your own desires or ever saying "I want" and "I prefer" and "I see it this way." It's easy—and far too common—for daughters to take a passive role in their own lives if they are accustomed to reacting to their mothers and forgetting, or never discovering, what it's like to have an identity of their own.

To individuate, the key task of adulthood, is to become your own woman, and it can't happen if you've never allowed yourself the freedom of venturing toward what you want, expressing your talents, and liking what you like. Within your own boundaries, you are in control. But until now, you've behaved as though that were a fantasy, an impossibility. Let me assure you that it's not—once you untangle your mother's life from your own.

Creating Boundaries

Boundary-setting is a four-step process that requires both inner and outer work, and a great deal of courage, for which I promise you'll be well rewarded.

Step One: Decide What You Want

If your boundaries have been weak, take time to think carefully about the kind of behavior you'll allow going forward in your relationship with your mother. What makes you feel invaded, diminished, belittled, powerless, or erased? Where do you want to draw the line between what you are willing and not willing to do in response to her requests? What is and isn't okay for you?

You have the right to determine what's allowable when you're together. Is it okay with you if your mother starts to restyle your hair? Does she need to call before coming over? Can she call late at night if it's not an emergency? When she's at your house, is it all right if she picks up a letter on the counter and looks at it, or goes through your drawers or refrigerator? Do you mind if she borrows things without asking? Reorganizes your closets? Reads the texts on your cell phone? You get to set the rules within your own "borders," and they'll be as distinct and individual as you are.

Remember that you *always* have the right to be treated with respect, and to protest unfair treatment or criticism. It's vital to reinforce those rights with boundaries. It's never okay for your mother to yell, threaten, or verbally abuse you. You have a right to ask her not to criticize you, your friends, or your family. You can ask her to withhold unasked-for advice. You don't have to accept any more blame or responsibility for her problems and upsets.

If you find yourself wavering or feeling unsteady as you prepare to start letting your mother know what you want, check back to your personal Bill of Rights. You are an adult with options and choices, and you can move a giant step closer to claiming your freedom by making a list of your mother's unloving behavior that you'd like to protect yourself from. The space you put between yourself and that behavior is a boundary. It's not necessary or advisable for you to put all your boundaries in place at once—you'll probably do it over time—but it's essential for you to be clear about what you want.

Step Two: Use Position Statements to Express Your Wishes to Your Mother

A boundary known only to you is meaningless. It becomes real when you clearly inform your mother of the new ground rules for your relationship with her, and then tell her when she's crossed the line and made you feel uncomfortable, or made a request, demand, or assumption you don't want to go along with.

You set boundaries with position statements. These are clear,

direct, nondefensive expressions of what you want. Position statements begin with words like:

- I am no longer willing to . . .
- I am willing to . . .
- It is no longer acceptable for you to . . .
- It's not okay that you . . .
- I need you to . . .

They sound like this:

- "Mom, I'm no longer willing to listen to your complaints about
  Dad. I need you to have those conversations with someone
  other than me."
- "It's not okay that you criticize my husband."
- "It's no longer acceptable for you to drink in my house or around
  my children."
- "I'm no longer willing to continue spending every Sunday with
  you. I am willing to make a monthly date, but I need you to
  make other plans most Sundays."

You don't need to apologize, explain, rationalize, or plead. Keep
your position statement on point: What's okay? What's not? That's
what you want to communicate.

Step Three: Plan Your Responses, Then Use Them

In a perfect world, you'd state your preferences, your mother
would give you a hug, say, "I didn't realize that was bothering you.
Of course I'll change!" and you'd have a wonderful relationship
from then on. But real life with an unloving mother is a good bit
messier than that. Your boundary-setting will come as a shock to
your mother. She's rarely or never seen you be assertive in this way,

and she's probably assumed you never would be. She may well be threatened by your newfound power, and she's certain to push back.

That's why it's so important to be prepared for what she's likely to say, and to practice the responses you'll give, just as you did when you were learning nondefensive communication. This time, you'll concentrate primarily on staying focused on your position without being drawn into arguments, explanations, or critiques of your own behavior. Here are a few examples (you'll see more later in this chapter):

WHEN SHE SAYS: "Why are you doing this now? It never seemed to bother you before."
YOU SAY: "It bothered me a lot, Mom. I just never had the guts to say anything about it before. I'm no longer willing to accept . . ."

WHEN SHE SAYS: "What's gotten into you?"
YOU SAY: "Courage and clarity, Mom. I'm no longer willing to overlook, excuse, pardon, or accept unacceptable behavior."

WHEN SHE SAYS: "Who put you up to this?"
YOU SAY: "It's all me, Mom. I've been giving this a great deal of thought, and I'm no longer willing to accept the status quo/ accept things the way they are."

Step Four: Decide on Reasonable Consequences

Your mother may resist or refuse to honor the boundary you've set, so you'll need to have a plan for what you'll say—and do—if her response to you is negative. I'm not talking about punishment; I want you to focus on taking yourself away from the harmful behavior. When you set a boundary, you've deemed particular behavior unacceptable to you. To protect yourself from the effects of that behavior, and to demonstrate your seriousness both to your mother and yourself, you'll need to put distance between yourself and the behavior. How do you do that when she pushes back? First, restate your position statement to make your boundary clear. Then, if it's still not honored, you can:

- Leave.
- Ask your mother to go.
- Hang up the phone.
- Set limits on your contact with her.

There are many other possible responses, so think about what makes you feel the most protected from the behavior that bothers you. The object of distancing yourself from your mother's unloving behavior if she doesn't honor your boundaries is not to retaliate, hurt, or shame her. Rather, it's to act in your own best interests.

Deciding what to do in advance, letting your mother know your plan, and making an unbreakable commitment to yourself to follow through will all ensure that your communication with her is clear, and that your actions will match your words.

You can't change your mother's behavior—only she can do that—but you do have the power to change the relationship by changing your own actions.

Lauren: Practicing Position Statements

After her success with nondefensive communication, Lauren was eager to take a next step and set some ground rules with her mother.

LAUREN: "It's such a small thing, saying something like, 'I'm sorry you're upset' instead of apologizing or caving in. But it works. The nondefensive stuff has made me feel a lot stronger. But

Mom's still expecting me to call every night, and instead of calming her down after I've 'missed' a call, I just need to get on with it and tell her that I'm not going to be on her schedule anymore."

That's where position statements come in, I told her. I explained the basic concept and asked her what kind of phone relationship would feel comfortable for her.

LAUREN: "I really don't want to talk to her more than two or three times a week. That would feel great. Maybe ultimately I could even decrease it. . . . And it doesn't have to be a long conversation. I just want to know she's okay, and let her know I am. But I just can't bear the thought of spending hours on the phone, with her grilling me about my personal life and making plans for me."

SUSAN: "How are you going to tell her that?"

LAUREN: "I can say something like, 'Mom, before we hang up, I need to talk to you. I want to let you know I'm not going to call you every day anymore. I have a life of my own. I'm happy to call you a couple of times a week, but it's no longer okay that you want all the details of my personal life during these calls. Also, it's no longer okay for you to call me at work. You can leave a message on my cell, but it's not okay for you to call me on my office line during the day.' But I just know that if I don't call her every day, she's going to keep calling and badgering me. That's what she always does. I just know that she's going to flip out and say I don't love her anymore."

Enmeshed mothers do what works for them, and Lauren had taught her mother that she would get her way by simply marching over her daughter's boundaries. Every time Lauren accepted that, she was reinforcing her mother's behavior. Now, to teach her mother new behavior, she would have to stand firm. Once she was clear about her boundary—two or three calls a week, no calls at work—she would have to tell her mother, and *show* her, that she was serious about it. "What will you do if your mother starts calling after you've told her the new ground rules?" I asked her.

LAUREN: "I don't want to cave as I've always done. . . . It will be hard, but if she calls me at work, I'll tell my secretary not to put her through, or cut the conversation and let her know I'm going to hang up."

SUSAN: "So you'll want to tell her that as well. 'Mom, I need you to respect my wishes and not call me between our twice weekly calls. And if you do call me at work, you'll leave me no choice but to ask my secretary not to put you through unless it's an emergency. This is serious, Mom.'"

LAUREN: "But she's going to go nuts. I can just hear her now saying, 'What have I done to deserve this? Why are you punishing me?'"

It's common for many mothers to cast themselves as the victim when you begin to resist their unhealthy behavior. That's a powerful form of manipulation. You don't have to accept it, I told Lauren. One effective response is: "Just calm down. There are some new ground rules in place and you need to hear them and take them seriously."

You'll be far more empowered if you don't respond to your mother's specific questions or accusations and instead stay focused on the limits and boundaries you are putting in place. It doesn't matter if you sound like a broken record. Your mother may be so off balance, shocked, or bewildered that she doesn't fully take in what you're saying at first. Just continue to set your limits, tell her very clearly what the new rules are, and spell out what you'll do if she disrespects or violates them.

"Believe me, I understand your reluctance to stand your ground and get your mother to take you seriously," I told Lauren. "I know that you feel really uncomfortable setting limits because it sounds harsh and unfeeling to you. And I respect that you don't want to hurt your mother's feelings. But you still need to take care of yourself. And if you don't, nothing will change."

LAUREN: "I know. Then I guess there's nothing to do but . . . do it."

I advised her to write out a short script for what she would say, and to memorize it, along with a few of the responses we'd talked about. As she'd learned with nondefensive communication, there's nothing like practice to make new words feel comfortable and natural.

Lauren was smiling broadly when she came in for our next session.

LAUREN: "I can't believe it, but I did it, Susan. It wasn't perfect—
I was nervous and I felt guilty as hell, but I did it. . . . The
hardest part was when she started crying and reminding me
of how she'd been there for me when I was younger. My heart
was pounding out of my chest, but I told her, 'Mom, we're
talking about the present now. We're two adults, and our
relationship isn't healthy—we're too enmeshed. I'm not saying
that you're totally responsible for this unhealthy relationship
we've had. In many ways, I've allowed it to happen. I do enjoy
being with you and talking to you, but not with the frequency
you want. I care a lot about you, but that doesn't mean I'm
going to continue this ritual. It's not okay for me, and it doesn't
feel good. I'm sorry you're upset.'
    "It was all I could do not to say anything more when there
was a long silence at the other end and she was sniffling
and crying. I felt awful. Finally she said, 'Don't you love me

anymore?' I said, 'Mom, don't be silly. Of course I love you. But I need you to respect my wishes. I need to go now. I'll talk to you in a few days.'

"At first she kept calling back, and I didn't answer the phone, so she left messages on my voice mail. She finally quit calling, though she sounded pretty upset in the messages I *did* listen to. But I'm here, I'm still standing, so I guess that means I'm surviving my guilt. It's been two weeks now with me being the only one who makes the call, and she's calming down. I'm really starting to feel a lot less smothered, and a lot less guilty. And I like her a lot better when I don't resent her so much."

I told Lauren how proud I was of her and how proud she should be of herself. Then I asked her to take her right hand, pat her left shoulder, and while she was doing that to say to herself: "You did really good, kid." She gave me a quizzical look, but she did it. And then she smiled and said she wanted to do it again.

With that first boundary, Lauren was establishing a precedent. It was the beginning of a process—with one position statement you can't expect an immediate 180-degree change in your mother. But by changing your behavior, you're creating an atmosphere in which she will either change hers slowly or demonstrate to you that she's unwilling to change. It will take time for you to find out, so be patient. This is new behavior for you and for her.

Karen: Responding to a Mother Who Goes Ballistic

Some mothers, like Lauren's, will resist your position statements at first but gradually begin to honor your boundaries. Others, though, may react badly, with yelling, threats, or extreme anger.

That's what Karen expected. She had been working with me to find new ways to respond to a volatile, controlling mother who had been highly critical and verbally abusive when Karen announced her engagement to her boyfriend, Daniel. (You can see our earlier sessions in the chapter on controlling mothers.)

Karen had come to see just how much her mother's control had poisoned her life, and she knew it was time to change from being reactive to being proactive, which would mean setting firm limits on her mother's behavior. But, like many daughters, she was scared to death to try.

KAREN: "In my mind I know how to do the position statements, but I don't know what she'll do. . . ."

"We have a lot of evidence of what your mother is like, and the way she'll respond if you challenge her," I told Karen. "She's insulted you and Daniel, tried to bully you, and threatened to withhold any support for your wedding. So we know that if you tell her you're not going to accept being treated that way any longer, she's not suddenly going to change her stripes and say, 'You're right. Let's sit down and plan the wedding, sweetheart.' It's just not going to happen. But we can work on strategies to deal with whatever she does."

KAREN: "Well, part of me is going, 'What's the point, then?' She's just going to blow up, it'll be really ugly, and I'll probably feel beaten down the way I always do."
SUSAN: "The important thing is that you'll be protecting yourself with firm boundaries. You'll be letting her know she can no longer control you. You can't let fear run your life. Is your fear more powerful than your commitment to yourself to have a better life? Is your fear more powerful than your love for Daniel? Is it more important than your self-respect?"

KAREN: "No, I really want to do this, and I know I have to. . . .
But I've tried standing up to her before and it never worked."

I assured Karen that she was far more focused and clear, now that she'd gone through the three steps of deciding what she wanted, practicing position statements, and deciding what she would do if her mother didn't honor her request. The script she prepared sounded like this:

"Mom, it's no longer acceptable for you to berate and insult Daniel, and it's no longer acceptable for you to insult me, either. From here on out, the subject of the wedding, and Daniel, is off-limits. It's not okay for you to tell me who I can marry. If the subject comes up, I'll end the conversation. If you insult me, I'll end the conversation or leave." It was just five clear, very powerful sentences.

Now she could prepare for the fireworks she was sure she would face. It was possible that her mother would surprise her and be reasoned and reasonable—but it wasn't likely.

First, I told Karen, she should determine the conditions for the boundary-setting conversation. Rather than letting anxiety build as she waited for the next unpleasant scene to unfold when she would challenge her mother's behavior, she could take charge and decide when she was going to initiate the conversation. And she could decide whether to have it in person or over the phone.

If you believe your mother will be abusive, there's no need to have a conversation in person or in a situation that makes it hard for you to leave if you need to. I suggest using the phone. It lets you say what you need to say in a space that feels safe to you, with your script in front of you, if that makes you feel more confident.

"Arm yourself with nondefensive responses," I advised Karen. With a volatile mother, it's especially important not to fan the flames. Don't argue or get defensive. Don't get into a yelling match. Don't engage in an exchange of blame or insults. Don't get sidetracked. Keep it simple and direct, and stay on message.

If you have an unloving mother like Karen's, she is likely to project blame onto you almost automatically, and she may attack you for daring to suggest there's anything wrong with her behavior. Karen was braced for a barrage.

KAREN: "I know exactly what she'll say: 'How dare you talk to me like that? Who the hell do you think you are! You're shaming the whole family by getting involved with this immigrant. He's beneath you. You'd better take a good look at yourself, young lady!' I'll be lucky to get a word in edgewise."

SUSAN: "You don't have to take insults and abuse—no one does. You can say, 'If you continue to insult me, I'll end this conversation.' You can also interrupt her by saying, 'Stop right there,' or 'That subject is off-limits,' or 'Unless you calm down, I'm going to hang up.' There's nothing chaining you to the telephone, and nothing chaining you to the chair if you're sitting across from her. You're an adult. Nothing will ever change until you change the way you characteristically behave with her.

"Until now, your mother has had all the power. But now you are telling and *showing* her that that has to stop."

Summoning her courage, Karen called her mother, and she reported back in our next session.

KAREN: "It was one of the hardest things I've ever done, Susan. My stomach was totally in knots. Just dialing the phone took a lot. And it was about as bad as I expected. I told her I wouldn't tolerate her criticism of Daniel and me, and that I wouldn't talk to her anymore about the wedding. That really put a stake in the heart of my crazy fantasy of us shopping together for a wedding dress. . . . She just roared at me, yelling about how awful I was and how much she despises Daniel. I held the phone away from my ear for a second, then I said, 'Stop right there. Screaming and berating me isn't going to work anymore.' She went silent for a second. I said, 'Your choice is to have a civil conversation with me or no conversation at all. Those are the only options, Mother. I need you to stop assaulting me. And I need you to understand that who I fall in love with and marry is my business, not yours.'

"All she could say was, 'Fine. Go ahead and ruin your life, see if I care.' And she slammed down the phone. It didn't feel like a total victory, exactly, and I was churning inside, but I was so relieved. I could look Daniel in the eye and not feel ashamed. I could really feel my integrity."

Taking steps to put distance between yourself and verbal abuse the way Karen did will probably leave you upset and shaky. It's not easy to face your mother's anger. But along with any inner turmoil you feel, I promise you that you'll feel pride, too, a sense of "Oh my God—look what I just did!"

"You now have the tools to protect yourself, and you can do it again, if need be," I told Karen. "Walls didn't cave in on you. The sky didn't fall. Breathe deeply, and do whatever you need to calm yourself down after a difficult exchange with your mother. Soak in a hot bath and remind yourself how courageous you've been. You *can* take care of yourself. You've tapped the warrior woman inside. She's always been there, waiting for you to find her."

It's Not Easy, Do It Anyway

This is the moment when you decide: "I'm either going to go on with the status quo, letting my relationship with my mother continue to erode my well-being, or I'm going to change."

There's no easy way to do this. You just have to do it, a step at a time. You may feel guilty or afraid, but promise yourself that you will tolerate that guilt, and any other discomfort you feel, in the service of becoming a healthy woman. That is the most important pledge you can make to yourself.

Chapter 12

Deciding What Kind of Relationship You Want Now

❦

"I finally feel like an adult."

Y ou've changed.

You're stronger now for having embraced your rights and taken steps to protect yourself. You've set boundaries with your mother and shifted your habitual responses to her behavior. You've reclaimed your integrity.

As you continue to use your new behavioral strategies, you'll find out if your mother is truly willing to respect your limits, boundaries, and wishes. It may take time for her to accept that you're serious, but each time you tell her what you're willing and not willing to do, and then hold firm, you're demonstrating that this is the way things are going to be—the new normal.

As that reality sinks in, some mothers, especially those who are not extremely critical, controlling, or narcissistic, realize that the stakes are high, and that if they want to preserve a relationship with you, they'll need to begin respecting you as an adult and equal. Others feel attacked and can't bear the thought that they may be in the wrong, so they defend themselves by *intensifying* their unloving behavior rather than examining and trying to change it.

You have four options for going forward—and maintaining the old status quo is not one of them. You can:

1. Continue using your assertive and nondefensive strategies to protect yourself and keep her unloving behavior in check. In some instances, that's all you'll need to do.
2. Negotiate for a better relationship. That requires spelling out what you want yet again, then coming to an agreement about what both of you will do, and monitoring her behavior and your own to prevent backsliding. Negotiation generally comes into play in more complex situations that can't be handled by gradually putting boundaries and limits in place.
3. Have what I call a tea party relationship. You maintain contact but keep it superficial, and actively protect yourself by closing off opportunities for your mother to see or criticize your vulnerabilities.
4. Cut off from your mother entirely.

We'll explore each of these options in the pages that follow, the first three in this chapter and the final option in the next. Remember that you are in control of this process—you set the terms of the relationship, and *you* decide what will serve you best.

Option 1: Use Your New Skills to Reinforce the New Normal

Lauren, who had used position statements to give herself more breathing room in her relationship with her enmeshing mother, was optimistic when she came to see me after a few weeks of keeping her new boundaries intact.

LAUREN: "I can hardly believe it. Our old phone check-in time
of five P.M. comes and goes, and sometimes I don't even think
about it. I actually kind of look forward to chatting with Mom
a few times a week. I'm not saying everything is magically
perfect, but things are a lot better. I have to watch myself to
keep us from going backward, because I know Mom would still
like us to be joined at the hip. But I'm getting better at letting
her know I can't do everything with her.

"I was planning a little dinner party, just some friends, and
when I mentioned it to Mom, she was really upset about not
being invited. I *almost* said, 'Oh, all right. Just come.' But then
I thought, 'No, that's crazy. I don't want her there.' And that's
pretty much what I said. I never used to be honest with her,
but this time I told her, 'Mom, this is the kind of thing I've been
talking to you about. There are times when I just want to be
with my friends.' I didn't apologize, not even when she went
back to the old 'You don't love me anymore.' It was hard. I said,
'That's silly, Mom.' And then I said, 'I love you, Mom. I've got
to go.' That was it."

Every time you express a truth that isn't cruel or abusive,
you grow. Lauren had stopped being the dutiful, guilt-bound
daughter—*"Yes, Mother, of course you can come"*—and was fast
becoming an empowered woman—*"This is what I want."* I asked
how she'd handle situations like the ones that had come up in
the past, when her mother would buy tickets to a concert without
checking with her and expect her to drop everything to go.

LAUREN: "I've thought about that. And I'm prepared to say, 'I
appreciate your generosity, but that won't work for me. I have
other plans.' As long as I remember how good it feels not
to resent her all the time, and how good it feels to actually
be honest with her, I think I can do okay. It occurred to me

that I used to lie to her all the time because I thought I'd hurt her feelings if I didn't. How can you hope to be close to someone that way? I think we have a chance of having a better relationship now."

I wish I could say that it's common for mothers to respond as Lauren's mother did, but many don't. And in that case, you'll need to take further steps.

Option 2: Negotiate for a Better Relationship

For many, even most, of the mothers you've seen in this book, a daughter's day-to-day assertiveness will have little impact.

Controllers and narcissists, especially, may be deaf to your position statements and go on as if nothing has changed. Or they may respect your boundaries for a while and promise that they've changed for good, only to revert to their old behavior. (Narcissists, especially, pride themselves on looking good, and it's not unusual for them to go along for a time simply to burnish their own image, and look as if they're really trying.) In cases like that, you'll need to have a more formal negotiation with your mother about your wants and needs rather than trying to make incremental shifts, issue by issue.

You'll also want to negotiate for a better relationship if you know that only a swift and significant change in your mother's behavior can give you any chance of healing. That's true when:

- Your mother has an untreated addiction or condition such as depression that prevents your relationship from improving.
- A crisis requires you to change the destructive aspects of your relationship immediately.
- Abuse was part of your history with her, and you need to see how willing she is to take responsibility for either her complicity

in the abuse or having been an active abuser herself. If she's not willing to take responsibility, it is destructive to your well-being to have a relationship with her at all.

In all of these instances, you'll need to tell your mother the critical issues you need to resolve and let her know what you'd like her to do, what *you're* willing to do, and what will happen if the situation does not change. This negotiation can protect you from your mother's unloving behavior and give you the degree of contact with her that you believe is healthiest for you.

Because the issues involved are so significant, and they may be complex and contentious, it's essential to prepare thoroughly for your negotiation so you'll have the inner clarity and confidence to communicate with your mother calmly and nondefensively. Sometimes, as you'll see below, you may feel comfortable enough, based on the experience you've had using position statements, to negotiate on your own. But if your anxiety level is off the charts, it's important to have the support of a therapist. If your family history involves physical or sexual abuse, the support of a therapist is *mandatory*. You need an advocate; it's unwise to take this on by yourself.

Allison: "I'm not willing to be your mother any longer."

Allison, who had grown up in a role reversal with her helpless, depressed mother, saw clearly how her caretaking tendencies with men grew from having been put in the role of being her mother's caretaker and marriage counselor.

ALLISON: "A light went on this week when my mother called me after another fight with my father. She was really upset. Apparently Dad came home from work to find her in her room, with no dinner happening, and he lost it. He started slamming doors and drawers, and then he emptied all the pots and pans out of the kitchen cabinet, yelling and swearing, 'You haven't even started!' I know it must've been awful.

"Mom started the usual litany of 'What am I going to do? What am I going to do? I don't know how much more of this I can take.' The first thing that went through my mind was 'I can fix this. I'm stronger now, I'm healthier. . . . I know exactly what she should do.' And then I caught myself. I took a deep breath and I said, 'Mom, you've been taking Dad's abuse ever since you were married. You're perfectly capable of figuring out what to do and making a change. But, instead, all you've been doing is shifting your unhappiness to me, and quite honestly, I'm not willing to take it anymore.'

"She started to cry, and I felt like a criminal. I almost called her right back to tell her, for the millionth time, that she ought to leave Dad. But then I thought, 'What am I thinking? I've been doing this my whole life, and it's never done any good.' You know how they say one definition of insanity is doing the same thing again and again and expecting a different result. . . . What does that say about me?"

I assured Allison she was far from being insane—and she'd done a good job of protecting herself. She was just becoming ever more aware of the way she had been programmed to fix, to solve, to take care of, to be the grown-up. And those automatic responses had to end.

ALLISON: "I know, Susan. I just can't keep trying to be her mother. She has to take care of herself, get some treatment. I can't do it for her."

But Allison *could* insist that her mother get the help she needed, and I told her she could make that a condition of continuing their relationship. Until the illness, whether it's depression or an addiction, is under control, you're in a relationship with the illness, not the person behind it. A decision about the role your mother can play in your life can only come after she's gotten help. And if she won't, there's no way for any involvement with her to be healthy for you.

You must not betray yourself by continuing to rescue someone who will not get treatment. This is essential.

If your mother refuses to get help, remember:

- Detachment isn't betrayal.
- Saving yourself isn't betrayal.
- Setting conditions for a healthier relationship is not betrayal.

They are healthy, adult responses to a toxic situation.

I asked Allison to rehearse with me exactly what she wanted to say, and the words poured out.

ALLISON: "I want to say, Mom, I'm no longer willing to be your mother. I'm not willing to solve your life problems for you. . . . I want you to know that I'm happy to talk to you on the phone, but when you start to moan and complain about your marriage and your life, I'm going to change the subject or say I have to go. I want you to be prepared for that. . . . Now listen carefully because this is important. The main condition for my continuing to have a relationship with you is that you see your doctor about your depression. I'm willing to go with you, but you must get treatment to get your depression under control. Will you do that?"

Allison was excited when she phoned me a few days later.

ALLISON: "I'm so relieved, Susan. It worked. I went over to Mom's house in the late afternoon before Dad got home and found her there in an old housedress with no makeup on, watching TV. I said, 'Mom, I need to talk to you. I want you to go wash up, put on some makeup, and get dressed, and then we can talk.' I made us some coffee in the kitchen, and then I told her I couldn't live her life for her—she had to take care of herself, and especially deal with her depression. She said, 'I know it's a problem, but I don't know what to do.'

"I said, 'I want to help you through this, Mom, but you have to keep your part of the bargain.' She squeezed my hand and said, 'I'll do it, honey. I know I lean too much on you.' Then she actually said thank you. I didn't expect that. I don't know why I never had the guts to have the conversation. I don't know what I thought she'd do. I actually feel hopeful."

There was a realistic basis for that hope. Her mother had made the small, but hugely significant, shift from being totally helpless to saying she'd take some responsibility for herself. For the first time, she agreed to get professional help.

Stacy: "Things have to change immediately."

Sometimes your mother seems to have so much power in the relationship that the consequences of a rift with her feel dire, or the pressure to come to an agreement with her is freighted with fears and concerns.

That was Stacy's situation. She had an urgent need to shift her relationship with her enmeshing mother, who lived a few doors down the street and had gotten into the habit of spending all her free time at her daughter's home. Stacy's husband, Brent, had given Stacy an ultimatum, saying he couldn't "be married to two women—you have to do something NOW," and the clock was ticking. (You saw my earlier sessions with Stacy in the chapter on enmeshing mothers.)

Because Brent and Stacy had relied heavily on the financial support of Stacy's mother—she had bought the house they were living in, allowing them to rent for a nominal amount as Brent built his business, and she babysat their young children after school—the idea of negotiating with her mother activated deep survival fears for Stacy.

STACY: "It's been going okay, setting limits with Mom—I told her it wasn't okay for her to poke around in our mail, and she actually stopped. But to tell her to . . . basically leave us alone unless we invite her over, which is what I know I need to do . . . I'm so scared about doing that. But if I don't, I'll lose Brent. I start getting heart palpitations every time I think about doing it."

I told Stacy that the best way to ease those fears was to channel her imagination and energy, which she'd been pouring into worst-case scenarios, into practicing what she needed to say to her mother. Being prepared—writing out a script and saying the words aloud until they feel natural—is the best way of gaining confidence as you learn to put your needs into words.

SUSAN: "I know this is difficult, so let's see if I can help you come up with some things you'll feel comfortable saying. Since I'm not emotionally involved in the situation, it will be easier for me. Your mother has to know that she can no longer assume she has the same rights in your house as she has in her own. She isn't respecting your separateness, your adulthood, or the fact that you have a partner. And the only way to let her know that is to tell her. You might want to take notes so you'll have a script that you can study and memorize. That way, you'll have the words you need to say if you get anxious and upset.

"You could say, 'Mom, I'm very grateful for what you've done for us, but this arrangement just isn't working out. It's very destructive to my marriage. In Brent's words, he didn't expect to be married to two people—you and me. I didn't expect to see us living as the Three Musketeers, either. So I've been giving this a lot of thought, and I'm going to give you some ideas of what is acceptable and what is no longer acceptable.

"'I know this is going to hurt you, but I'm caught in the middle, and it's not a very happy place for me to be. I care about you, but we've got to disentangle our lives. We're too fused, and it's not healthy for any of us.

"'I haven't wanted to bring this up because I didn't want to hurt your feelings, but now I have this situation where Brent says he's ready to leave if things don't change. I can see how much harm it's done to us for me to hide how unhappy I've been with this arrangement. We need to be able to have a life of our own that's separate from you. You've never accepted the fact that I'm not in this world to keep you company. And in many ways it will be better for you when you do. You're a bright, professional woman, and all you're doing is being a babysitter.'"

I paused to point out to Stacy that she could be kind and respectful without getting into apologies and defenses. This early part of the negotiation was to state the facts without accusations. The next step would be to lay out her position statements, which would consist of: "I'm willing to do this . . . I'm not willing to do this . . . I'd like you to do this."

I told her that those position statements might sound like this: "We'll be happy to have you over for dinner when it's convenient for us, which may mean once a week, or every other week. But not every night, Mom. Brent and I have had precious little time alone since we were married. That has to change.

"It's no longer acceptable for you to criticize my husband and come over whenever you want and stay as long as you want. If there's a special program on TV that we all want to watch together, fine, but after it's over, you need to leave. We are not the Three Musketeers.

"I really appreciate all you've done to take care of the kids. But they're both in school longer now, and I'll either cut my hours at Brent's company so I can be here when the kids get home or I'll find someone to watch them. I'm also going to need to get the house key back from you. I've been too dependent on you, and you've been too dependent on me. I'm a healthy adult woman with a family of my own, and I don't need you to hover over me anymore. You need to have activities and friends of your own."

Stacy gave herself a deadline of having the talk with her mother before her next appointment with me, and she told me about her negotiation when we met the next week.

STACY: "Mom was really indignant at first. She said something like, 'I can't believe you're doing this. What's happened to you? After all I've done for you and all we've been through, I can't believe that husband of yours is forcing you to do this.' My heart was pounding, but I kept taking deep breaths and I said, 'It's not acceptable that you talk to me that way, Mom. I'm an adult, and this is coming from me, not from Brent. This is what *I* want.'

"She just looked at me. All the color left her face, and she looked so pained and deflated, as if I'd punched her in the stomach. Then she said, 'What have I done? All I've ever tried to do is help.' That was so hard. I just repeated, 'This is what I want, Mom. Things have to change.' Finally, she said, 'I can't bear the thought of not having you in my life.' I said, 'That's not what we're talking about, Mom. We're talking about a mature adult relationship between an adult daughter and an adult mother. I have a partner now. I know you wanted to have me as a partner all your life, but it's not going to happen anymore.'

"She looked really surprised. Then she said, 'Does this mean you don't want me to see the kids?' I said, 'Of course not.' It was the hardest thing I've ever done, but I said, 'Mom, I think it would be best if you gave me back the house key now.' And both of us were crying as she pulled it off her key chain and handed it to me. She just said, 'I should go now.' It was awful. I have never felt guiltier in my life. . . . But something funny happened. I knew I had to tell Brent right away, and called him. He said, 'Honey, I love you so much. I'm so proud of you. I knew you could do this.' He sounded so relieved. And for the first time, I felt like we'd be okay on our own, as our own family. I know Mom is upset. But I feel strong enough now to handle it. After all, I'm married to Brent, not to her. And this could be just the push she needs to start making a life for herself that's less focused on us."

Now Stacy is putting into practice behavior that is bound to strengthen her, and her marriage.

Remember that it's not possible to negotiate at all with a person who becomes enraged, irrational, verbally abusive, or aggressively critical. If your mother has responded combatively to the work you've done so far, she has closed off the negotiation option. If she *is* able to listen to you, but you're worried about being able to deliver your message calmly and clearly, you can also send her a letter that follows the format I gave Stacy, first presenting the facts without accusation or apology and then listing your position statements. When in doubt, seek out the support of a therapist who can help you.

Kathy: "You need to acknowledge your role in my abuse."

Kathy had reached the stage of her therapy when we were looking toward winding up. She had worked hard and done all her assignments and exercises. Her motivation to get better, and to heal the depression that was the legacy of her sexual abuse as a child, had paid off. She was quite different from the withdrawn, worried woman I first saw, and both her husband and children benefited from the changes in her.

KATHY: "There's one thing I still need to do. I've stayed in touch with my mother by mail and an occasional phone call, but we never deal with the elephant in the living room. Now that I'm so much stronger I'd like to test the water and try one last time to see if we can salvage something out of the debris. I know she's hurting, too, and has for years. We've never talked about the role she played in letting my father abuse me. Can I ask her to come here? Would you see us together?"

I told Kathy I thought it was a great idea, but I cautioned her to keep her expectations low. Her mother might refuse to come in— in which case she would be giving Kathy all the information she needed about whether a closer relationship was possible. Or she might come in and be very defensive and closed. Kathy said she felt able to handle those possibilities and wanted to go ahead.

She decided she would write to her mother, who lived in the Midwest, and before she mailed her letter, she made a copy for me. Here is an excerpt:

Dear Mom—
I'm almost ready to finish up therapy, and I'm at a point now where you could help me a great deal. I need you to come in for at least one session so we can see if we can salvage some of the good stuff that's between us and get rid of a lot of the disturbing stuff so we can make our relationship better. I love you, Mom. I want us both to be well. Please come to California and we'll have a "come to Jesus" meeting with my therapist. She'll help us get back on track so we

can both stop being victims. I need you to do that. And you
need it more than you know. It'll give you a chance to do
something positive for me. I'll be waiting to hear from you.

With hope,
Kathy

Kathy's mother, Andrea, e-mailed her daughter and said she
would come the following week, and when she arrived, Kathy
drove her from the airport to my office.

Andrea was a well-groomed, nice-looking woman in her six-
ties, with a heavy air of sadness about her. She told me she was
extremely anxious about what was going to happen, but she would
do anything to help her daughter. I assured her this was not going
to be a "Let's beat up on Andrea" session, and I thanked her for
coming in. Kathy and I had spent a session going over what she
wanted to say until she felt prepared, and I asked her to start.

She talked openly about the sexual abuse and how angry she'd
been for years at being unprotected, and she told her mother how
much she needed her to acknowledge what had happened. She was
brave, forthright, and clear in her expression of what she wanted
and needed from her mother. This is part of what she said:

KATHY: "In a couple of unmailed letters I wrote in therapy, I
expressed more anger at you for my grotesque childhood than
I've ever expressed before. When you are abused continuously,
you get very angry. Because there's nothing to justify being
emotionally or physically cruel to an innocent child. Just
because I haven't really expressed this anger toward you
doesn't mean it still isn't there.

"The truth is that during and after all the horrible things
my father did to me, you protected him. That makes you his
accomplice. Somehow, I feel that you've always known that
and felt guilty for it. I would feel guilty, too, if I were you. He's

the criminal, but he never paid for his crimes because you covered up and lied and stuck your head in the sand to protect him. And to protect yourself, too. Because if the situation were never confronted, you wouldn't have to go through the embarrassment, as you put it, of answering for him.

"You were more concerned about being embarrassed than about protecting me. You never got to know me as a normal little girl. You didn't make it possible for me to be one. You missed out on a lot, Mom, all because fear took up so much of your life. And I missed out on a normal childhood altogether.

"I want you to take responsibility for that, Mom. I have nothing but contempt for my father, but I think if you can take responsibility for your part, then you and I can go on from here. Because despite everything, I still love you, Mom. I want us both to get well."

Andrea listened quietly, head down, and hands clutched in her lap. I told her I knew how hard this was for her and asked her what she wanted to say to her daughter.

ANDREA: "No matter what I say it will be inadequate after all the harm that has been done to you. At the time I thought I was protecting you as best I knew how, but I was so full of fear, and the disbelief that he could really be doing what he was doing, the uncertainty. I didn't know what I should do. So I didn't do anything. I . . . let him hurt you. The years came and went. . . . I never did claim to be a strong person. . . .

"I realize to my own disgust how I could be so easily manipulated. I guess my thinking was that it was so horrible and no one should know—how everyone would laugh. . . . But all I did was harm you. I can't forgive myself for that. I'm so sorry, honey. I don't have words for how sorry I am. Maybe I'm not making much sense. . . . I love you, Kathy. It's a mother's duty to protect her child, and I was so lacking. Nothing that happened to you was your fault. Please believe me when I say that. I don't know if you can forgive me, honey. Letting him hurt you is the worst thing I've ever done, and I live with the guilt every day. It has left me with very little self-worth and very low self-respect. I'm so sorry. . . ."

By this time both Kathy and Andrea were crying, and I had teared up as well. Andrea had come through for her daughter at last by not simply saying "I was a bad mother" but by being specific about what she'd done, and taking responsibility for it. It was important for her, and especially for her daughter, that she reassured Kathy that she wasn't responsible for the abuse she'd experienced. Andrea's willingness to say those words helped to further free Kathy from lifelong doubts, and diminished the guilt she'd been carrying so long. Now it would be possible for the two of them to establish a relationship based on honesty. And they'd have to take it one small step at a time.

Over the years, I've often been surprised by which mothers come through for their daughters in sessions like Kathy's. Sometimes it's the mothers you'd least expect even to show up because they've been so consumed with guilt about their actions and so unwilling to go into that dark place to tell the truth. You won't know until you ask.

If you were physically or sexually abused and want to see if a relationship with your mother is possible, it's essential to do it with the help of a therapist. Your mother, like Kathy's, has to take responsibility for what she did, and she has to get some therapy or go into therapy with you for a few sessions.

I want you to know that you can absolutely overcome the dark legacy of sexual abuse. I've helped literally hundreds of women (and men) with that trauma. With good, active therapy and compassionate support, it's possible for you, too, no matter what your mother does. You may wind up with a closer relationship with her, and you very well may not. What's important is that you will have done everything you can to have a compassionate, loving relationship with yourself.

Option 3: The Tea Party Relationship

If your mother has been resisting the way you've changed, or demonstrated that she's not interested in meeting you even partway, one way to maintain contact without damaging your well-being is with what I call a tea party relationship. It's entirely superficial. You don't bring up anything that can give your mother an opportunity to hurt or criticize you, and you never make yourself vulnerable.

This is an option many daughters choose because it allows them to protect themselves while maintaining some kind of contact, and to orchestrate their interactions with their mothers with much less anguish than in the past. Sometimes, they decide, a safe, artificial relationship is better than no relationship at all.

Jan, whose mother alternately supported her blossoming acting career and chipped away at her confidence by criticizing, competing with her, and negatively comparing her with other women, had worked hard to connect with the "good mother" who sometimes appeared. (Our earlier sessions are in the narcissist chapter.) But no amount of boundary-setting could keep the inevitable criticism and comparisons away.

JAN: Things have gotten a little better, but Mom will always be
   Mom, I guess. She's always got that verbal knife ready, and I

never know when she's going to stick it in me. When I tell her I'm no longer willing to accept her criticism, she looks at me, nods her head, says, 'I understand'—and then she goes right back and does it again. I show her a great photo shoot from our set and all she can say is, 'That's nice, dear. And by the way, your hair would photograph so much better if it were a few shades lighter.' Honestly, her mouth should be registered as a lethal weapon. But that's just the way she is, and I don't think she's ever going to change.

"I know you're going to say I have to stay away from her, but I'm just not ready to cut her out of my life altogether. She's my mother. And I still have so many good memories of her. She can be pretty nice when she decides to be."

I told Jan I understood her desire to stay in contact with her mother. But I strongly suggested that she pull back and limit how much she involved her in her life. "Don't talk about parts you're up for, or your hopes and dreams," I told her. "Don't invite her to anything professional that you're doing—because you know how much she'll compete for attention. Have a chatty, surface relationship where you talk about movies and books and the weather, and don't open yourself up to the kind of clobbering you know you're going to get if she needs to make herself better by one-upping you. You're in a field that's very tough on self-esteem, and the last thing you need is your mother making you feel insecure on top of that."

You don't need a therapist to help you navigate a "tea party relationship," but you do need to keep your guard up and be fast on your feet. You'll have to actively change the subject when it gets too close to sensitive territory—the topics you care most about, and the truths of your life that you know she'll reflexively criticize rather than support. A relationship like this is a lot like fencing. She thrusts, you deflect, and you do a kind of choreographed dance designed to keep her at a safe distance.

If Jan's mother called to say, "How are you? Are you working? Had any auditions?" it would be Jan's job to change the subject by saying, "I'm fine, Mom. Did you see that great movie on HBO last night?" Because Jan's mother was so narcissistic and competitive, it would be easy to get her to talk about herself.

If you choose this option, expect to be bombarded by "Why?" questions as you close off large swaths of your life from her:

- Why are you acting this way?
- Why can't we talk anymore?
- Why won't you share your life with me?
- Why are you so closed?

You can say:

- I'm just working through some things on my own right now. Mom, what have *you* been up to?
- I'm not ready to talk about that right now. Maybe later.
- I'm great, Mom. I want to hear about *you*.

To avoid your mother's probes and deprive her of targets, you'll need to weave and dodge and creatively bring up new topics, even when you're feeling annoyed or off balance.

JAN: "I can see how tricky this can be. I thought talking about a movie I saw would be safe and neutral, but as soon as I brought it up, Mom said, 'You should've been in that one. How come your lousy agent didn't have you up for the part of the daughter?' It brought me up short for a minute, but I regrouped and said, 'By the way, I just read a book I think you'd like.' I have to be on my toes all the time."

This isn't how most of us would prefer to behave with someone who's supposed to love us, and it may seem very phony to you. But if you are either not ready or not willing to cut the emotional umbilical cord with an actively unloving mother who is not combative or overtly abusive, this is certainly an option. Because you're being actively self-protective, it empowers you, and that changes the status quo. Don't think that you are copping out if this option feels best to you. Many women are not willing to take the step of detaching from actively unloving mothers, and they're relieved to find a middle ground that keeps their integrity intact. Sometimes, that can be your healthiest choice.

Chapter 13

The Most Difficult Decision

᭦

"It's come down to a choice between my mother and my well-being."

No one can magically give a mother the desire to have a better relationship with her daughter. Some mothers' defenses, and the unloving behavior that comes with them, are stronger than any maternal feelings. When that happens, some women have the skills and willingness to maintain a distant, superficial relationship. But others find that a tea party relationship can't insulate them enough from their mothers' control or criticism or enmeshment. That leaves one very difficult option: breaking off contact.

No one comes to this decision lightly. Through all the work they do on themselves and all the skills they learn, daughters watch for any sign that their mothers are responding positively. It's wrenching for them to realize that it's not going to happen and that the only way to end the destructive patterns that are eroding their lives is to end the relationship.

Deciding to break off contact, and then following through, is one of the hardest things a woman will ever do. It may be *the* hardest. But for those who have exhausted their options, and themselves, it's often the step that frees them to have the healthy, rewarding lives they long for. In this chapter, I'll show you how I helped my client

Karen navigate this difficult option. I don't recommend taking this step without support, and if you are considering it, I urge you to seek out a therapist who will actively guide you, keep you focused, and validate you through the process of both cutting off and rebuilding a life without your mother in it.

Breaking Off: When Everything Else Has Failed

Breaking off is a last resort. It comes after women have run out of ways to give their mothers the benefit of the doubt and can no longer cling to the impossible hope of a miraculous "happily ever after"—or even a nontoxic next encounter.

Karen, whose mother actively opposed her plan to marry her fiancé, Daniel, worked hard to set limits on her mother's insults and attacks on the relationship. But her mother had been deaf to her position statements and her firm, nondefensive requests for change.

KAREN: "I just had another call from Mom. I really can't listen to her go on about Daniel anymore, so I've been keeping her at arm's length. I've tried everything. I thought we could have a tea party relationship, but all she ever wants to talk about is how Daniel is ruining my life. Nothing is working, Susan. Nothing. She calls my position statements 'psychobabble.' But today was the worst. I did what you said last week and wrote her a letter to try and negotiate for a better relationship. I told her that if we were going to have any kind of relationship, the topic of Daniel was off-limits, and I also said we couldn't stay in contact if she didn't get some therapy.

"Well, I got my answer. She called me this morning and immediately started screaming at me. She said, '*I'm* not sick. *You're* the one who's sick. Everything would be fine if not for this intruder coming into our lives.' I cut her off before she could go on, but it was awful. What am I going to do?"

Even when you've braced yourself for a negative reaction when you spell out your conditions for the relationship, it's deeply unsettling, and even shocking, to get a response from your mother that says, in essence, "No. I'm not willing to bend. Not even when I know how much I'm hurting you." Any denial about how little she's able to respect your needs and wishes vanishes in the harsh light of that response.

I asked Karen what she wanted to do.

KAREN: "I didn't actually believe that it would come to this, Susan. . . . But if she doesn't change—and she hasn't changed a bit—it could ruin my relationship with Daniel. . . . I'm at the end of my rope. She's controlled and humiliated me so often, and I just don't think I can be around her at all anymore."

A daughter like Karen, who comes to the conclusion that she can't have the life she wants with her mother in it, is faced with a choice between her mother and her emotional well-being. It's imperative that she choose the latter.

Once she's made that pivotal decision, the steps she takes next must clearly demonstrate her strength, independence, and ability to follow through on what she says—not just to her mother, but also to herself. She must hold fast to the knowledge that she's a strong, independent woman, not a helpless child, and can survive in the world without her mother. That means putting aside all the tenacious "what-ifs" and "if onlys" and "if I'm just good enough she'll have to love me" longings, and all the fantasies of what might have been. "It's time for you to release those damaging fictions once and for all," I told Karen. "They never served you well."

Telling Her It's Over

The best way to tell your mother about your decision to break off contact with her, I told her, is through a short, direct letter. This is not a time to recap old hurts, ask questions, or demand apologies. The job of this letter is to tell her, simply and clearly, that it's no longer possible to have a relationship with her. The letter should be brief and nondefensive. It shouldn't take much more than a paragraph or two.

I told Karen to structure the letter like this:

"Mom, after much thought I have decided that it is in my best interest to not have any further contact with you. That means no phone calls, no letters, no e-mails, and no coming over. I am not going to spend any more time with you. Please respect my wishes."

It's important to avoid double messages that leave the door open a crack, I told her. Her mother wouldn't take her letter seriously unless it was unambiguous.

I told Karen that it's not a good idea to try to communicate this message in person. An action like this wouldn't be necessary if her mother had been at all receptive to her requests, or even willing to listen. Daughters need to stay calm and focused on what they need to say, and the most effective way to do that is in writing. I ask my clients to write their letters by hand and mail them instead of using e-mail. The sight of her daughter's handwriting will underline to a mother that the words aren't coming from an anonymous machine.

My last bit of advice for Karen was not to do this by phone—it would be too easy for her mother to hang up or bombard her without letting her finish.

Karen's letter closely followed the above format, but of course she struggled to get the words down.

KAREN: "At first, I had things in there like, 'All I want is for you to meet me halfway, I know we could have a good relationship if only you could see how good Daniel is.' I pleaded and wished and, to tell you the truth, I really grieved. Because when I reread that first draft, it really hit me how little I've gotten from all my years of wishing and pleading." (Her eyes welled up, and she took a deep breath before continuing.)

"So I remembered what you said about keeping it short and to the point, and got on with just telling her. It felt right to spell out what I meant by cutting off contact, because honestly, I don't know if she'd get it otherwise. I know she doesn't believe I'd ever have it in me to do this."

Karen read her letter to me, and I told her that she'd done a fine job. I reassured her that what she was feeling was expected. Daughters report being flooded with many emotions as they wrote their cutoff letter—sadness, disappointment, fear about consequences, self-doubt, a terrible sense of loss, and most of all, overwhelming guilt about what they were doing. Karen hadn't let those feelings stop her as she focused on what she needed to do, and wrote her letter. She'd now have to keep facing down those emotional demons as she dealt with the inevitable fallout.

KAREN: "So . . . I just pop the letter in the mail and wait for the bomb to explode?"

SUSAN: "You focus on your life with Daniel and the people who really love you, you plan your wedding, and feel what it's like to live without those daily doses of negativity from your mother."

KAREN: "I'm looking forward to that. But this seems so . . . final. I know I've done everything I can—I *know* that, but it's going to be rough for a while. I know my family—the relatives on Mom's side—are just going to flip out. My guilt went through the roof just now when I read the letter to you. It's not so much Mom I'm worried about, it's everyone else. . . ."

Exorcising the Guilt

Before we could work through Karen's fears about how her family would react, and come up with strategies for handling other people's responses, I needed to help her calm her guilt about the enormous step she was taking. Many daughters believe they don't have the right to cut off from their mothers—after all, if it's a taboo to say anything bad about your mother, it can seem unthinkable to end the relationship with her, even when staying in contact is seriously undermining their well-being. Daughters feel incredible guilt—for challenging the status quo, for taking an action that shakes the foundation of their concept of family, for daring to do what's best for them instead of continuing to sacrifice themselves to other people's expectations. Perhaps most of all, they feel guilty for being the one to say "enough" and cut the ties that bound them to a mother who could not love them and wouldn't begin to try.

One of the best techniques I know for dealing with that mountain of guilt is to pull it into the light and challenge it. I asked Karen to do that by finding an image of a "monster" that she could use to represent her guilt and anxiety, and then talking to it and letting it know that it could no longer run her life.

I suggested that she search for an image online and print it out or tear one from a magazine. Karen found hers in *National Geographic*, a sea monster guarding the corner of an old map.

She set it in front of her, stared at it for a minute, and began to speak:

KAREN (to her guilt): "I don't fully understand how you came to be such a big part of my life, but I am here to tell you to get out. I no longer need you, I don't want you, and I'm not going to satisfy you. You made me do things that violated my own self-respect and integrity.

"You made me afraid. I was afraid of consequences, of doing what I wanted and knowing who I really am. It's your fault that I feel so tormented about doing what I need to do to have the life I want. I've spent so much time satisfying you and making you happy and being what you wanted me to be. Those days are over!

"I'm the only person whose approval I need. I need to learn to be happy with who I am and what I want to be. I get to choose who I let into my life and who I keep out of it, and how I want my life to play out. *I* am in control of this, not you.

"I'm not going to let you demonize me for doing what's best for me.

"Good riddance!"

KAREN: "Wow, that surprised me, all that power and determination."

A sense of power often surges when daughters sit down and tell their guilt they're not willing to let it steer their lives anymore. In doing that, they're also informing their unconscious mind that they will no longer let their emotional demons stand in the way.

Strategies for Handling the Reactions of Family and Friends

Karen had been sustained by a loving aunt and cousins, and her greatest fear was that she'd lose them. Many women worry not only about the repercussions of breaking off with their mothers but also about being frozen out of the rest of the family. Upsetting the balance of a whole family system by changing your own behavior can be frightening.

KAREN: "I don't know what to do, what to say, how to tell them. . . ."

I told Karen she wouldn't have to worry about telling her extended family—her mother would almost certainly let everyone know what had happened. Unloving mothers are likely to sound the alarm, rallying support for themselves to oppose what many are likely to describe as their daughters' "sickness" or "outrageous behavior."

KAREN: "God, the family is going to come down on me like a boulder. I don't know how my aunt will take it, and I know there are people who are really going to go ballistic."

I advised her to be prepared for a variety of reactions—including positive ones from unexpected quarters. "You don't know who will do what," I told her. "But remember that people who really love you will support you as you do what's best for yourself. And you can use all your nondefensive skills to deal with the ones who don't."

Daughters often face family members who call up as advocates for their mothers and demand apologies. Relatives may blast a daughter with criticism. In a religious family, they may invoke their tradition's version of "Honor thy mother." They may blame the daughter for breaking up the family or say things like, "You're killing your mother. She's crying herself to sleep every night."

I remind my clients that they don't have to let themselves be bombarded or patiently sit still for a tirade. They've had more than enough of that. I advise them to use the communication skills they've learned and turn to reliable nondefensive responses such as, "I'm sure you see it that way" and "You're entitled to your opinion." I also suggest using statements like this:

- This is between my mother and me.
- I don't choose to have this conversation.

- This is my decision and it's not negotiable.
- This topic is off-limits. If you want to talk to me, we'll have to talk about something else.
- I know you're concerned, but I don't want to discuss this.

While a daughter needn't discuss her decision with every aunt, uncle, and cousin, it *is* important that she speak individually to members of her immediate family—her father, if he's in her life, and her siblings if she has any—to let them know that she's taken this step to protect her emotional health. You can't control their reactions, I tell my clients, but you can urge them not to take sides.

KAREN: "What about family stuff like birthday parties and Christmas? Do I have to call and see if Mom's going to be there?"

SUSAN: "What would you think about not going at all? I know that might be hard. But I think it would be very difficult for you to be in same room with your mother. That could activate all the old patterns you've worked so hard to diminish.

"Remember, you're trying to build a new life. Make it clean. There may not be as many people around you when you do this, but the ones who are left will be good for you instead of destructive."

Karen decided she wanted to make a little ceremony of dropping her letter to her mother in the mailbox and asked her fiancé, Daniel, to go with her.

KAREN: "He put his arm around me and told me again how much he loves me and how proud he was of what I'm doing. I cried. I thought I was over the grief, but I guess that will take a while. I feel so close to Daniel now, though. It really helps."

In the weeks after she sent her letter, Karen did face some of the angry phone calls and encounters with family members that she'd expected. But not everyone was incredulous or upset.

KAREN: "My aunt Meg, my mother's sister, the one I was so worried about? I couldn't believe what she said when I had lunch with her and told her what I was doing. She put her arm around me and said, 'I understand perfectly, honey. Your mother's always been a bitch.' I had to laugh, and I hadn't laughed in a while. I know Meg will be there for me. She always has been. I can't say this has been easy, but I've got the comfort of knowing she really loves me. Meg even offered to stand up with me at the wedding."

There's no sudden "happily ever after" once a daughter breaks off contact with her mother. Many of my clients tell me that they feel great relief and pride, and almost everyone wrestles with self-doubt and guilt for a while. It's not uncommon at all to bounce between highs and lows. After a particularly rough encounter with her sister, who blasted her for "letting that man break up our family and devastate Mother," Karen was shaken.

KAREN: "I know I did the right thing. But it's hard to see everyone so upset. What if they're right and I just made the biggest mistake of my life? When I hear myself say that, I know it's not true, but I'm so up and down about this. It's hard sometimes to be the one who divorced her own mother."

I reassured Karen that the doubts would ease. "You have to remember that you're taking good care of your self-respect and your integrity, and that will pay off in a big way as you go along," I told her. "Do you really want to go back to the way things were, and live with your mother's constant criticism of you and Daniel, just so you can placate your sister and other family members? You're powerless to change your mother, but you're doing a great job of changing yourself, and that's all you can do. All this second-guessing will ease up. Time is your best friend right now. You're not going to feel great all at once, but it will happen, day by day."

I reminded her that family isn't determined solely by blood— and that she was discovering her family of choice, the people who loved and respected and valued her enough to stay in her life now.

KAREN: "You know, you're right. Daniel's family has been so
    wonderful to me, it's like they've adopted me."
SUSAN: "See, you do have a family. And you and Daniel will have
    a family, too."

It takes time for life to settle into its new shape after daughters have delivered the news to their mother. I always urge my clients to have a strong support system in place—not just a therapist, but also true friends and supportive family members who can remind them of the importance of sticking to their decision, even in the face of extreme pressure or hostility. "You will grieve, and you will have to keep facing down guilt and uncertainty," I told Karen. "But little by little, the pain will disappear. And you'll feel the roots of a new, healthier life spreading beneath you."

Chapter 14

Old, Sick, or Alone:
The Suddenly Dependent Mother

꠸

"I have to be there for her.
After all, she *is* still my mother."

All the work that goes toward healing the wounds of growing up without adequate mothering brings daughters immense benefits: less negativity, guilt, and fear. A quieting of the compulsion to be a people pleaser. A life piloted by their own desires instead of anyone else's. A circle of family and friends who genuinely love and respect them. A blossoming sense of confidence and courage. It takes daily effort to keep old patterns from reasserting themselves. But once that work becomes a new way of navigating life, it is impossible to imagine going back to the painful constraints of the way they lived before.

One particular set of circumstances, though, can easily throw anyone off balance: dealing with a mother who is suddenly ill, infirm, or alone. A serious crisis can reopen old wounds, unravel a daughter's careful, self-protective decisions about the relationship, and reactivate not just old, unhealthy patterns of behavior but also the longings that so often underlie them.

But because life isn't static, and challenges arise, I want you to read this short chapter, think about it, and keep it in reserve—you may well need it at some point. It will help you keep your life on course, and protect your own well-being if it is challenged in ways that become unhealthy.

Deborah: "Mom's Got Cancer"

It's difficult for *any* adult daughter to define her responsibilities to a mother who's facing the challenges of aging, or whose life has suddenly been turned upside down. But for a daughter who has carefully taken steps to change her relationship with an unloving mother, and perhaps even to cut off contact with her, it's excruciating. When an unloving mother suddenly has a broken hip or a life-threatening illness, or calls tearfully to say, "Your father is dying," what happens to the careful boundaries you've put in place to protect yourself?

It's easy to regress. All the guilt about setting limits, and all the lingering desire for love and approval, come to the surface as a daughter faces a mother in need. A daughter may have done significant work to repair her life and to become more confident and independent. But even when she is feeling relatively strong as a separate person, the reappearance of her mother, this time in troubled circumstances, can reactivate old patterns. Every daughter hopes that a crisis will turn into a blessing—that a brush with mortality or an immersion in grief will become a catalyst that dissolves the worst of her mother's unloving behavior and brings her closer. It can happen. But because that outcome is far from certain, I always advise clients in this situation to temper their hope with caution, but also leave open the possibility that this may be a time when they can forge a new connection with their mothers.

Deborah, who had been physically and verbally abused by her mother as a child, had come for treatment when she feared that her anger at one of her children had come close to spilling into dangerous territory. We worked hard to deactivate the shame, grief, anger, and destructive programming from her childhood, and in our time together, she had decided that she could only have a superficial relationship with her mother. "I don't want to completely deprive the kids of their grandma—they only see the good side of her that I never had. So we have her over for dinner occasionally, and the conversation is all about the kids. That's it. She and I don't really talk much," Deborah told me in an e-mail six months after we'd ended our work together. "I'm so much happier now. I can't undo the past, but the present is okay."

But a couple of years later, she got news from her mother that changed everything. "Can you squeeze me in today?" Deborah asked when she called me. "I have to see you." She stopped by my office on the way home from work.

DEBORAH: "I got this call from my mother last night. She just found out she's got breast cancer. It looks like stage II, I'm not sure—they're talking about lymph nodes and chemotherapy and . . . Oh my God, Susan. I was on the Internet all night reading and trying to see what the options are and what we need to do. My brain just won't shut off, but I'm so scared for her I can't think straight. I have this terrible feeling that they're going to find more . . . that she's going to die."

"I'm terribly sorry, Deborah," I told her. "That's of course a very real possibility, and you need to be prepared for that. But let's slow down and take things one step at a time. Making a list of your questions and talking to your mother's doctors so you're operating on solid information is one good step. Taking advantage of the counseling and support the hospital offers is another.

"Knowing you, I know that your first impulse is to drop everything and be there twenty-four/seven for your mother. But you've got a husband, three young kids, and a thriving business, and you can't turn your back on any one of those things. So let's work together here to figure out how much you can realistically give your mother without wiping yourself out."

No matter how badly she was mistreated, it's difficult for a daughter to do anything other than be pulled back into her mother's world when her mother is suffering. In the first days and weeks of an unfolding crisis, she may well be immersed in questions, arrangements, and intense emotions—hers, her mother's, her family's. She may be reeling; that's to be expected. And with her mother's needs looming large, she probably will have a hard time concentrating on her everyday life.

But from the very beginning, it's vital for a daughter to put her own needs on the endless to-do list. Who might help her with tasks? Who can share the responsibility for helping meet her mother's new needs? Whom can *she* turn to for emotional support? No matter how much a daughter believes that she is the only person who can deal effectively with the problem and no matter how much she's *encouraged* to believe that, there is always help available. She has to make a priority of finding it.

That is easier said than done, and it's not uncommon for women to crash first before realizing they have other options. For Deborah, the days after her mother's diagnosis were a blur of activity, and it was much easier for her to run herself ragged than to rest and seek support. She was exhausted when she came in a month into her mother's treatment.

DEBORAH: "Mom's doing all right, and I have to be thankful for that. They think they got all the cancer with the surgery, but now she has to go through a lot of chemo, and it's grueling. It's torture for her, Susan. So I can't really complain about taking care of her."

I asked her to tell me about what she'd been doing, and she seemed to trip into high gear.

DEBORAH: "Well, I've added a few more things to my day, you
    could say. I get the kids off to school, and on my way in to work
    I check in on Mom, see if she needs anything, if she's eating.
    Then I try to get some work done between calls to doctors and
    research I've been doing on clinical trials, and if she needs a
    ride to the hospital I take her and bring her back. I pick up
    the kids, make dinner, make food for Mom, run by to check
    in on her, try to get back to my computer to post updates for
    relatives. . . . And then I see if I can make a dent in the work I
    didn't get done during the day. I haven't figured out how sleep
    fits into that. Or my own lunch, some days."

That had to be taking a heavy toll on her, I said. "It's making me exhausted just listening to you."

DEBORAH: "I can't complain. I'm glad I can be there for her.
    She lights up when she sees me, and I know my presence is
    making this whole awful process a little easier. She's told me
    that, Susan. She said she doesn't want a stranger in the house
    looking after her—she wants *me*. And for the first time I
    can ever remember, she said she loved me." (Deborah's eyes
    welled up with tears.) "I've been waiting for that my whole
    life."
SUSAN: "I know you have. It's good for you to take it in as deeply
    as you can. If your mother is offering warmth or closeness or
    gratitude now, embrace it, and take it a day at a time."
DEBORAH: "That's what I'm trying to do. I know it might not
    last, but right now I feel like I have the mother I always
    wanted."

For some daughters a long-yearned-for change in their mothers is an answered prayer, and it often takes great effort for them to remember that the lives they've built for themselves still need attention, too.

"How are you holding up?" I asked. "Tell me what's happening with your family and your business."

DEBORAH: "That part's not so great. I think I'm close to losing a client I thought I could count on for next year because we've been late delivering proposals to them. The kids are getting a little whiny about not seeing me. Jerry is trying to be supportive, but he's really resentful. He says things like, 'Your mother never did anything but hurt you, and *now* all of a sudden she's being nice because she's sick. Why do you have to be the one to rush in and take care of her every need?' He thinks my sister should fly in and help, and Mom should hire someone to help her at home.

"I know I'm getting worn down. I was knocked out with a bad cold last week, and I can see that I can't keep this up. I guess I'm not Superwoman after all. . . . But she's my mother, and this could be my last chance to have her."

Of course Deborah wasn't the only one who could give her mother what she needed. As I told her, "*You're allowed to have a life*, and you may need to tell your mother you'll engage a caregiver or that you'll spend a reasonable amount of time with her but not bear the entire responsibility for her care. You can't go on like this."

DEBORAH: "I feel so overwhelmed, Susan. I hardly know how to start thinking about this. I don't want to lose her."

"And you don't want to lose yourself, either," I said. "Probably the most important thing you can do," I told Deborah, "is to be proactive in finding support for yourself. This is one of the toughest challenges you will ever face. On one side you have your mother, who's very ill. And on the other you have your own business and family. You can't be the full-time caregiver for anyone and still have the energy and vitality you need to keep going. If you go down the tubes, how does that help your mother?

"Let's figure out together how you can lighten the load for yourself," I told her. "You can cut back and enlist help from other resources. You can hire help, or if money is an issue you can organize your mother's friends to help, and people from her temple. Your time will be so much better spent finding support rather than running around."

We made a list of questions for her to answer. What kind of resources did her mother have? Who could take her mother to at least some of her chemotherapy and doctors' appointments? Who could help with food and care at home? I gave her the assignment of researching some options on the Internet and checking with the support groups at her mother's hospital. Doing this turns on the rational part of your brain, I told her, and writing out "What do I need/what does she need/who can help" lists, then researching, is a useful process for turning whirling anxiety into the beginnings of solutions.

DEBORAH: "Jerry's been telling me the same thing, and I know he'll help me with this. I was so afraid that she'd get upset if I cut back any time with her."

Deborah ultimately found a meal service that specialized in delivering meals and snacks to cancer patients for a reasonable price, as well as a volunteer service that could take her mother to some of her chemotherapy sessions. Her mother had savings that could go toward the meal service, and when she was able to eat regular meals, Meals On Wheels was an option. Deborah also began talking to students from the community college who might do some driving and errands for her mother until she felt stronger.

There's no easy and permanent solution to a crisis like the one Deborah faced, but once she admitted that she couldn't—and didn't want to—bear the burden of her mother's care alone, she gained breathing room. And perspective.

DEBORAH: "I think the hardest part for me is realizing that I can't fix everything. I can't make her better, I can't do everything she needs, and I can't make her happy all the time. She hasn't been so happy with having more people involved, and those smiles and 'I love yous' aren't there so often. She's going through hell, and her moods are up and down. Last time I went by to see her she said, 'I had such a terrible night. You should've come. I hate having strangers here.' All I could say was, 'I know, Mom. I'm doing everything I can.' Sometimes I feel so guilty that I can't be everything to everybody. But I can only do what I can do."

"That's right," I told her. "And taking care of yourself isn't turning your back on your mother."

How Much Is Enough?

How does any daughter determine just where her responsibilities to her mother begin and end? It's a loaded topic, and in a crisis, you may well be surrounded by friends and family members who are certain they know what's right for you and your mother. But you are the only one who knows what you can handle, and what you need to do to preserve your own health and sanity. Your mother's illness or widowhood isn't an excuse for her to behave badly. It doesn't obligate you to tolerate having your life turned upside down, despite great pressure on you simply to go along with her requests and demands. Here, too, you'll need to stand up for yourself, difficult as that can be.

I don't suggest that you turn your back on an ill or suddenly needy mother, but in helping her, you'll have to decide what you can handle and then stand firm.

If she's ill, it may be that what you can do is talk to her doctors and help make decisions about her treatment, but not be involved in her day-to-day care at all.

You may realize that you have the time and willingness to spend a great deal of time with your mother for a month after she's widowed, then help her find other sources of solace and companionship.

Even in such extremely stressful situations, you must do what's healthy for *you*. As you did when you renegotiated your relationship with your mother the first time, you have to attend to your own needs and boundaries. Then, if you choose, you can offer your mother the best care and attention you can reasonably give.

Hold On to Your New Empowerment

It *is* possible to withstand the sometimes intense pressure to put someone else's concept of duty or their definition of your role as a daughter ahead of what you know is best for yourself. If you're faced with a mother in crisis and have trouble putting yourself in the picture according to expectations set by others, fall back on your assertiveness, your nondefensive communication, your boundary-setting. They'll give you the time and space to look out for your own interests and act on your inner wisdom.

And if you still feel torn or guilty, remember how much of your life you spent as the one for whom promises were rarely kept, the one whose needs rarely mattered. The neglected side of yourself is still there inside you, healing now because at last it sees you honoring it and all you can be. Remember that when you think your needs count for less than someone else's. Your well-being depends on it.

Coda:
Connecting, at Last,
with the Good Mother

ⵖ

I t's been a long, demanding journey. And in taking it, you have
joined a vast community of women who are reclaiming their
lives from the painful legacy of growing up with an unloving
mother. As you move ahead with a sound inner compass and a
whole new set of life skills for letting other people know what you
want, you can finally make your life reflect who you really are.

But if you're like many of my clients, you may still wonder: How
can I make up for the mothering I missed? How can I give others,
especially my children, healthy nurturing when I didn't experience
it myself?

The desire for good mothering never goes away. All the emotional work you've seen is process, not perfection, and regrets about
the love you missed may resurface, even after therapy. But the pain
will be manageable—it will feel more like a twinge than a stab.

And fortunately you can be soothed and nourished to an amazing degree by the mothering energy of others. There are many
people who can fill the role of good mother for a woman who never
had one—grandparents, other relatives, friends, lovers, anyone
who values and respects you. Every smile, kind word, and act of
appreciation from them can feed you.

There is also a good mother inside of every daughter, a wellspring of nurturing that can feed you and then flow through you to the important people in your life. You can connect with this source of warmth and caring in a number of ways: by observing mothers around you and refining your sense of what "good love" looks like; by remembering the times you were loved, and the people who have genuinely loved you; and by communicating directly with the wounded child inside you to give yourself the mothering you missed. As we wrap up our work together, let's look briefly at each of these healing choices.

Learn by Observing Good Mothers

Many daughters who've grown up without adequate mothering live with the paralyzing fear that if they have children, they'll somehow turn into the kind of mother they always swore they would never be. And if they do have children, and make mistakes with them— as every mother is bound to—they're certain that they're doomed to be just like their mothers. Daughters without children fear that they'll relate to their friends and lovers in the same destructive ways they learned from their mothers, becoming neglectful or critical or domineering or clingy.

I want to reassure you that you're very different from your mother. You have awareness and empathy, which she didn't. She may have spewed out terrible words and punishments without a thought to how wounding they would be. Or maybe she smothered or ignored or abused you. Whatever the behavior, she couldn't see beyond her own needs and impulses to care about the consequences for you. But from that came a very valuable gift: You know both intellectually and emotionally what every child and loved one deserves and should have. How sad that you had to learn it the hard way.

Your own nurturing instincts are good. You can trust them. But if you're feeling unsteady, you can build your confidence by observing mothers with their children.

Emily, who had been neglected by her mother, felt a longing for children that was matched only by her fears of "lacking the mother gene." In the time since I'd last seen her, she'd decided that she needed much more closeness than her increasingly distant boyfriend Josh could give her. After many attempts at salvaging the relationship, they'd finally decided to call it quits. About six months before she came to see me, she'd met a man through one of her friends, and now the relationship was becoming serious. "We're even beginning to talk about marriage and kids," she told me.

EMILY: "The big sticking point for me is that I just don't feel . . . qualified to be a mother. How can you give a child what you didn't have yourself? One thing I refuse to do is have a child and then screw up somehow. And given my history, I'll probably get everything all wrong."

I assured Emily that with all the work we had done, she didn't need to be afraid to have children. She had a keen sense of how she wanted to act as a mother, which she could depend on to guide her.

To help her see that for herself, I gave her the assignment of spending some time during the week watching friends, relatives, and strangers interact with their children. "Watch how a good mother behaves, and what the impatient, angry ones do," I told her. "The best mothers pay attention to their children, and at the playground, they're the ones waving back to their kids on the jungle gym rather than texting every minute. They're protective, but not smothering. Notice how they praise and encourage their children for trying, not necessarily for succeeding. And watch the way they use discipline. If their children misbehave, they'll take away privileges, but they won't assault their dignity or value. You won't have any trouble telling the good moms from the bad. Remember: If you can recognize loving behavior, you can emulate it."

Emily took my suggestion, and she reported back at our next session.

EMILY: "On Saturday morning I went to the playground in the park near where I live. There were a lot of moms. I sat on a bench by myself and watched the kids first. They were so great, so full of fun and energy. And so loud! And then I started watching the moms. There were lots who were hardly there. I could see how disappointed one little girl was when she got all the way to the top of the slide and was trying to get her mom's attention, but Mom wouldn't look up from her phone. But there were a couple of moms who seemed to be enjoying their kids as much as I was. It wasn't that they were with them every second. They were just like home base, and their kids could run over and get a hug before they ran off again for their next adventure. And the smiles between them—I really loved it. That's the kind of mom I'm going to be."

I encourage you to spend time in this way, soaking up what the give-and-take of healthy nurturing looks like. It will support you in any kind of relationship, whether you have young children, grown children, or no children in your life at all. Absorbing the mix of connection and freedom, attention and affirmation, in the bonds between good mothers and their children will reinforce what you now know about the workings of genuine love.

If you're considering becoming a parent, or want to improve your skills, I urge you to seek out some of the many, many resources available for mothers. Good examples and potential mentors are all around you, and whether you're asking questions on a mommy blog or hanging out with a friend's kids at the park, you can find opportunities to connect with the power of mothers' wisdom, and the burst of energy that children can bring into your life. If you're just imagining yourself as a mother, test the waters by being a loving "aunt" to a neighbor's daughter, or by helping chaperone some second graders on a field trip. And if you're already a mom, don't isolate yourself from other mothers. Plug into a mothers' network and risk being vulnerable with your questions and concerns. You don't have to struggle alone.

Remember the People Who Have Genuinely Loved You

Genuine love values and respects you. It embraces and encourages. It makes you feel safe and celebrated for who you are. Your mother may not have been able to give you that kind of loving, but you got tastes of it from other people all the same. And you can connect with that by going back to the times in your life when someone made you feel cherished. Closing your eyes and sitting with the memory can bring the feeling to life again. You can amplify that feeling with the Good Mother Exercise, a visualization that will quickly bring you in contact with the power of real love.

Here's how the exercise goes:

Sit in a quiet, comfortable place where you won't be interrupted and bring to mind one of the people in your life who acted as a good mother to you. It might be an aunt, a teacher, a grandmother—someone who treated you with kindness and respect, someone who cared about your emotional well-being and wanted to nurture it. Close your eyes and imagine that you're a little girl on a sandy beach with glittering waves washing gently onto the shore. Now picture your good mother coming toward you with a big smile and bright eyes that tell you how thrilled she is to see you. She runs toward you and wraps you up in her arms, and you bury your head in her shoulder. You feel safe, cared for. Stay in that place and those feelings as long as you like.

Now become that good mother yourself. Visualize holding that little girl—you—in your arms and say out loud, "I love you, sweetie. I treasure you. You're a wonderful child. I'm so glad you're my daughter. My life is so much richer because you're in it. I love you very much."

Those are the words of the mother we all deserved, and very few of us got.

Many of my clients feel very ripped off as they do this exercise for the first time. But I tell them to keep repeating it at home until the sadness diminishes. It will. The unconscious is a sponge, absorbing everything you send into it. And the more Good Mother Exercises you send it, the less room there will be in your head and in your heart for old, hurtful messages like "You're a terrible girl. You can't do anything right."

Your unconscious won't say, "That's just me talking to myself." It will absorb the experience. In essence, you're reparenting yourself, giving yourself the mothering you deserved so much.

Soothe the Wounded Child You Were

You can take your reparenting one step further by writing a letter to the wounded child inside. Like the Good Mother visualization, this letter taps the good mother energy that's always available to you and uses it to comfort the child you were. This letter speaks directly to old hurts, and it communicates in the words that the little girl inside you has waited a lifetime to hear.

In this letter, tell that child all the things you would have wanted your mother to say to you. Tell her that she is safe now and loved, and that you will always be there for her. It's very important to write this letter even if you don't have children and don't plan to have any. You still need to comfort and validate that part of your-self so you're free to love deeply and generously, without clinginess, desperation, or fear.

Here is a part of the letter that Emily wrote. It's a wonderful example of how reparenting the child that you once were nourishes both the little you and the adult you.

> Dear little Emily:
> I am so sorry that you were not treated very well when you were little. I am so sorry that your mom was not affectionate with you. I am sorry that you were never cuddled. I'm sorry that your mom never did fun stuff with you like reading a book together, going to lunch together, going to a movie together. If I could be your mom, I would start by tucking you in every night, giving you a kiss, and telling you how much I love you and how special you are to me. I wish I could have been your constant companion. I wish I could have been a soft chest for you to cry on and warm arms to rock you and to whisper, "There, there, I know you're so sad and angry. It's okay, my darling—cry it out."

The more you give that inner little girl the love she has always longed for, the more you will free up to give to your partner, friends, family, and, of course, your children. In this way you change not only yourself but also the world around you, and the lives of gener-ations to come. There's no need to fear that the love you've worked so hard to bring into your life is finite, that it will retreat or disap-pear. Love is like a homing pigeon—we send it out, but it always returns.

Real love, as you know so well now, doesn't make people feel unlovable or inadequate, or as though there is something terribly wrong with them. Love feels warm and safe. It makes your life better, not worse.

You are capable of that kind of connection. And as you learn to give yourself the mothering that your mother could not offer you, you are opening in yourself the ability to give—and receive—the tenderness and caring you've craved for so long. You have changed and grown. You can love.

Acknowledgments

I'm not a fan of long acknowledgments, but there are a few key people who were vitally important in bringing this book to fruition, and I want to thank them.

My gifted writing partner, Donna Frazier, as always, has been a source of enormous wisdom and strength. This is our fourth book together and not only are we still speaking but our relationship has grown closer over time.

My warrior agent, Joelle Delbourgo, has believed in me and my work from the time she was my editor for two of my previous books. She is also a very calming presence for this very emotional author.

My current editor, Gail Winston, brought her superb editorial skills and guidance to the project, and I am extremely grateful to her.

On a personal note, my wonderful daughter, Wendy, and her significant other, James McKay, constantly warm and enrich my life with their love, humor, and unflagging support.

And, finally, a deep thanks to the clients who gave so willingly of their stories and worked so courageously to heal their mother wounds.

Suggested Reading

Ackerman, Robert J. *Perfect Daughters: Adult Daughters of Alcoholics.* Deerfield Beach, FL: Health Communications, Inc., 2002.

Anderson, Susan. *Black Swan: The Twelve Lessons of Abandonment Recovery: Featuring the Allegory of the Little Girl on the Rock.* Huntington, NY: Rock Foundation Press, 1999.

Beattie, Melody. *Codependent No More: How to Stop Controlling Others and Start Caring For Yourself.* Center City, MN: Hazelden, 1986.

Brenner, Helene. *I Know I'm In There Somewhere: A Woman's Guide to Finding Her Inner Voice and Living a Life of Authenticity.* New York: Penguin, 2003.

Brown, Nina. *Children of the Self-Absorbed: A Grown-Up's Guide to Getting Over Narcissistic Parents.* Oakland, CA: New Harbinger Publications, Inc., 2008.

Collins, Bryn C. *Emotional Unavailability: Recognizing It, Understanding It, and Avoiding Its Trap.* New York: McGraw Hill, 1998.

Cori, Jasmin Lee. *The Emotionally Absent Mother: A Guide to Self-Healing and Getting the Love You Missed.* New York: The Experiment, 2010.

Engel, Beverly. *The Nice Girl Syndrome: Stop Being Manipulated and Abused—and Start Standing Up for Yourself.* Hoboken, NJ: John Wiley & Sons, 2010.

Fenchel, Gerd H., ed. *The Mother-Daughter Relationship: Echoes Through Time.* Northvale, NJ: Jason Aronson, Inc., 1998.

Forward, Susan, with Craig Buck. *Toxic Parents: Overcoming Their Hurtful Legacy and Reclaiming Your Life.* New York: Bantam, 1989.

Forward, Susan, with Donna Frazier. *Emotional Blackmail: When the People in Your Life Use Fear, Obligation, and Guilt to Manipulate You.* New York: HarperCollins, 1997.

Hotchkiss, Sandy. *Why Is It Always About You?: The Seven Deadly Sins of Narcissism*. New York: Free Press, 2003.

Lazarre, Jane. *The Mother Knot*. New York: McGraw-Hill, 1976.

Lerner, Harriet, PhD. *The Dance of Anger: A Woman's Guide to Changing the Patterns of Intimate Relationships*. New York: HarperCollins, 2005.

Love, Patricia, with Jo Robinson. *The Emotional Incest Syndrome: What to do When a Parent's Love Rules Your Life*. New York: Bantam, 1990.

Martinez-Lewi, Linda. *Freeing Yourself From the Narcissist in Your Life*. New York: Tarcher, 2008.

McBride, Karyl. *Will I Ever Be Good Enough?: Healing the Daughters of Narcissistic Mothers*. New York: Free Press, 2008.

Neuharth, Dan. *If You Had Controlling Parents: How to Make Peace With Your Past and Take Your Place in the World*. New York: Harper Perennial, 1999.

Secunda, Victoria. *When You and Your Mother Can't Be Friends: Resolving the Most Complicated Relationship of Your Life*. New York: Dell, 1990.

Solomon, Andrew. *The Noonday Demon: An Atlas of Depression*. New York: Scribner, 2002.

Spring, Janis Abrahms, PhD, with Michael Spring. *How Can I Forgive You?: The Courage to Forgive, the Freedom Not To*. New York: Perennial Currents, 2005.

Wegscheider-Cruse, Sharon. *Learning to Love Yourself: Finding Your Self-Worth*. Deerfield Beach, FL: Health Communications, Inc., 2012.

Index

About the Author

SUSAN FORWARD, PHD, is an internationally renowned therapist, lecturer, and author. Her books include the number one *New York Times* bestsellers *Men Who Hate Women and the Women Who Love Them* and *Toxic Parents*, as well as *Betrayal of Innocence, Obsessive Love, Emotional Blackmail, When Your Lover Is a Liar,* and *Toxic In-Laws.* In addition to her private practice, she has served widely as a therapist, instructor, and consultant in numerous Southern California psychiatric and medical facilities. She has appeared on over three hundred television and radio shows, and hosted her own nationally syndicated program on ABC talk radio for six years.

For counseling information you can reach Susan on her website at www.susanforward.com, or e-mail her at susanforward6@aol.com.